CELL CULTURE

CELL CULTURE

By

Dr. P.R. Yadav
Lecturer
Department of Zoology
D.A.V. College
Muzaffarnagar (U.P.)
(India)

&

Dr. Rajiv Tyagi
Department of Zoology
M.M. College
Modi Nagar (U.P.)
(India)

DISCOVERY PUBLISHING HOUSE PVT. LTD.
NEW DELHI-110 002

First Published - 2005

Reprinted - 2014

ISBN: 978-81-8356-019-1

Published by:

DISCOVERY PUBLISHING HOUSE PVT. LTD.
4383/4B, Ansari Road, Darya Ganj
New Delhi-110 002 (India)
Phone: +91-11-23279245; 23253475; 43596065
E-mail: discoverybooksindia@gmail.com
discoverypublishinghouse@gmail.com
web: www.discoverypublishinggroup.com

Printed at:
Infinity Imaging Systems
Delhi

1

ANIMAL TISSUE CULTURE

Every living organism is composed of one or more cells. Our understanding of cells dates back to 1665, when the British scientist, Robert Hooke described certain structures in a piece of cork as *cellulae* which is Latin for 'little rooms'. At about the same time the Dutch scientist, Antonie van Leeuwenhoek, was designing the first simple microscope which proved to be a vital tool in our understanding of cells. Systematic studies of microscopic anatomy were performed by Matthias Schleiden and Theodore Schwann in the 1830s and these led to the formulation of the 'cell theory' which has two basic tenets. Firstly it states that all organisms consist of one or more cells, and secondly that the cell is the basic unit of structure for all organisms. This theory was extended in 1855, when the German physiologist Rudolf Virchow stated that all cells can only arise by cell division from pre-existing cells. Since all life has a cellular basis, much research has been done in order to understand how cells work.

Multicellularity and Differentiation

One of the characteristics of most animals is that they are multicellular—in other words—they are composed of many cells. With this multicellularity comes the specialization of cells. In a multicellular organism, each cell does not have to carry out all the activities necessary for the life of the organism. Although most cells of any higher organism have many organelles and metabolic pathways in common, each cell is also unique in expressing some of these components to an enhanced degree in order to fulfil a specific function within the organism. Each cell type has its own role—to secrete a specific product, to contract, to transmit an electrical impulse, and so on. The result of this cellular specialization is that animals consist of a number of different types of cells—each with a characteristic size, shape, structure and function. Such cells are said to have differentiated. A vertebrate has more than 100 different types of cells. These cells associate in very organized patterns to perform specialized functions.

Many animal cells can, with special care, be induced to grow outside of their organ or tissue of origin. Isolated cells, tissues or organs can be grown in plastic dishes when they are kept at defined temperatures using an incubator and supplemented with a medium containing cell nutrients and growth factors. The *in vitro* cultivation of organs, tissues and cells is collectively known as *tissue culture*, and is used in many areas of science.

Historical Background

Tissue culture is not a new technique—in the scientific literature there are references to its use dating back to 1885. An embryologist, called Roux, was able to maintain the medullary plate of a chick embryo in warm saline for a few days. This was the first recorded example of successful explantation. In 1903, Jolly made detailed observations on *in vitro* cell survival and cell division using salamander leukocytes. In the early experiments, fragments of tissue were studied, and this gave rise to the name 'tissue culture'. Officially, the term tissue culture is used when cells are maintained *in vitro* for more than 24 hours.

Aseptic Technique

One of the main difficulties of animals tissue culture was keeping the cells free of contamination. Since bacterial cells grow faster than animal cells, contamination quickly leads to bacterial overgrowth. In 1913, Alexis Carrel applied *aseptic techniques* to tissue culture. He introduced the 'Carrel flask' which facilitated culture under aseptic conditions. To meet the nutritional requirements of cells, embryo extracts or animal blood serum were added to the cells. These were particularly vulnerable to contamination but the addition of the antibiotics, penicillin and streptomycin, to the cell culture medium from the 1940s onwards alleviated this problem. The problems of microbial contamination have also been greatly reduced by the use of laminar air flow cabinets, which minimize the possibility of contamination by air-borne microbes.

Cell Culture

Another significant advance was the use of trypsin (a proteolytic enzyme) by Rous and Jones in 1916 to free cells from tissue matrix.

It was subsequently used for the sub-culture of adherent cells and in the 1950s, the technique of trypsinization was exploited to produce homogeneous cell strains and this marked the start of animal cell culture techniques. Trypsinization is the term which applies to the treatment of cells by the proteolytic enzyme trypsin to change their adhesiveness. The term 'cell culture' refers to cultures derived from dispersed cells which have been taken from the original tissue. In cell culture the cells are no longer organized into tissues.

Modern Tissue Culture

Today, tissue culture techniques are considerably easier than they used to be, with standardized media and sophisticated incubation conditions. In the 1940s and 1950s, tissue culture media were developed and conditions were worked out which closely simulate the situation *in vivo*. In particular, the environment is regulated with regard to the temperature, osmotic pressure, pH, essential metabolites (such as carbohydrates, amino acids, vitamins, proteins and peptides), inorganic ions, hormones and extracellular matrix. Among the biological fluids that proved successful for culturing cells, serum is the most significant. 5-20% serum is usually added to media for optimal cell growth. Serum is an extremely complicated mixture of compounds which includes undefined components, and much work has gone towards creating a chemically defined alternative to serum.

Cell Types which Grow in Culture

For many years, most cell types were very difficult to grow in culture, but recent modifications in culture methods have allowed many specialized cells to be grown in culture. Literally thousands of different cell lines have been derived from animal tissues. The list of different cell types which can now be grown in culture includes connective tissue elements such as fibroblasts, skeletal tissue (bone and cartilage), skeletal, cardiac and smooth muscle, epithelial tissue (liver, lung, breast, skin, bladder and kidney), neural cells (glial cells and neurons, although neurons do not proliferate *in vitro*), endocrine cells (adrenal, pituitary, pancreatic islet cells), melanocytes and many different types of tumour cells. The development of these tissue culture techniques owes much to two major branches of medical research: cancer research and virology.

Cancer research

Human tumour cells were obvious subjects for research, since the discovery by George and Margaret Gey in 1952 that human tumour cells could give rise to continuous cell lines. The first human cell line to be grown continuously in the laboratory was the HeLa cell line, which was obtained from a malignant adenocarcinoma of the uterine cervix. This cell line opened up the possibility of characterizing the cells of malignant tumours *in vitro* and is still one of the most popular human cell lines for study. Later, non-malignant rodent cells in culture were used to analyze the effect of chemical carcinogens, viruses and other agents on normal growth.

Virology

In 1949, Enders *et al* reported that the poliomyelitis virus could be grown in human embryonic cells in culture which led to the production of the polio vaccine for mass vaccinations in 1954. The first human vaccines were produced in primary monkey kidney cells, but these were later found to carry a virus (called SV40) so they were not considered completely safe for the production of human vaccines. In 1962 a human cell line was developed and used for the production of the vaccine. Between 1954 and 1970 a wide range of human and veterinary vaccines were developed *in vitro* and licensed, including vaccines against measles, rabies, mumps, rubella and foot and mouth disease virus. Standard conditions were needed for the production and assay of viruses, and this led to the commercial supply of reliable media and sera, and greater control of contamination with antibiotics and clean air equipment.

The Applications of Tissue Culture

Monoclonal Antibodies

Since the 1950s a range of other products synthesized from animal cells have found commercial application. The next major development was the production of monoclonal antibodies. In 1975 Kohler and Milstein produced the first *hybridomas* capable of secreting a single, specific antibody. The generation of monoclonal antibodies is indispensable in many areas of biotechnology and is just one example of a routinely used production process which is completely dependent on tissue culture techniques.

Recombinant Proteins

Another important development has been the production of recombinant proteins from cultured animal cells. Large scale cultures of animal cells are becoming increasingly important in the production of a range of valuable products. In particular, much effort has been directed to the production of lymphokines, interferons, and hormones like human growth hormone. In 1986, human interferon γ (which is thought to have anti-tumour activity) produced from lymphoblastoid cells in culture was licensed as a therapeutic agent.

Reconstitution and Replacement of Damaged Tissues and Cells

In addition to using cells as factories to make viruses or proteins, there is also great interest in the cultured cells themselves as a product, for use in the reconstruction of damaged tissue or as a replacement of non-functional cells and tissues. *Tissue grafting* may be necessary when tissue has been lost, or has become dysfunctional for whatever reason. The missing tissue can often be replaced and in some cases even whole organs can be replaced if there is a source of healthy tissue. Unfortunately it is not always easy to find an appropriate source of tissues and organs. One source is from people who have died recently, but the tissue is recognized as foreign by the recipient's immune system, so there are problems with immune rejection which have to be dealt with. Another source is to use tissue from another site in the patient (autografting) and this is possible for skin grafts over relatively small areas, but is not possible for larger bum wounds or for most other types of tissue replacement. Cells cultured *in vitro* would be an obvious potential source of the missing function and this possibility has been examined for a number of diseases. Cultivation of the patient's own cells *in vitro*, in order to generate enough cells and/or to genetically manipulate the cells to replace a missing function, would overcome the problems of immune rejection and of scarce tissue sources.

Living tissue equivalents that survive upon transplantation in animals have been constructed from isolated cells of human skin, thyroid and blood vessels. We will now examine some possibilities for using tissue or cells raised *in vitro* as tissue replacements.

Vascular tissue

The latter is particularly important since any reconstituted tissue or organ will only survive and develop if it is in contact with a blood supply. A blood vessel model reconstituted from collagen and cultured vascular cells has been studied in depth with a view to therapeutic tissue reconstruction.

Liver

The liver is composed of cells with a remarkable capacity for regeneration and the isolation of lineages of parenchymal cells from the liver holds promise for the correction of organ-specific disease. In animal trials, researchers have shown that injection of new liver cells into a scaffolding of polymer sponge or beads which have been injected into the abdomen can take over some of the functions of a diseased liver prolonging the life of the animals for several months. The whole field of tissue reconstitution is expanding rapidly. We will illustrate this with some examples.

Parkinson's disease and neural grafts

Parkinsons's disease is a neurological disorder which causes tremor, rigidity and disturbances of posture. Chemical analysis of brain tissue involved in some types of Parkinsonism has revealed a marked depletion of a neurotransmitter called dopamine and this finding has led to attempts to introduce cells making the missing substances.

Recent developments in the treatment for Parkinson's disease include the replacement of cells in the brain with neural cells from another source by neural transplantation. The beginnings of this technique can be traced back to the last century, although the first neural grafts on humans were not performed until 1982. Studies indicate that these neural grafts are effective because they release certain chemicals which can be utilized by the host brain. There are two sources of tissue which have been used in clinical studies to date. Firstly, human embryonic neural tissue (from aborted human foetuses) has been grafted into the brains of patients with Parkinson's disease and this has met with some success. This approach is complicated by ethical controversy and the possibility of immune rejection of the tissue. Secondly, extracts of the patient's own adrenal medulla have been used in a process known as autografting. The adrenal glands are endocrine glands found just above the kidney and the logic behind this choice of tissue is that the adrenal medulla is derived from the neural crest and could, therefore, be expected to have some of the properties of neural cells. However, autografting of adult tissue has not been very successful.

One of the aims of researchers now is to use cultured cells to supply the missing chemicals to the brain. This will involve identifying the missing factors which are provided by the neural graft and to find a cell line which makes this factor, or to genetically engineer cultured cells to make the factor. Parkinson's disease is only one example of a disease which may one day be cured using cultured cells.

Immunosuppression therapy

Patients may be immunosuppressed for many reasons; for example because their immune systems have been suppressed by the drugs needed to treat cancer or to prevent graft rejection, or by diseases such as

AIDS. Such patients become particularly susceptible to infection by pathogens and one of the major threats, striking about 50% of bone marrow transplant recipients is the *cytomegalovirus* (CMV) which can cause fatal pneumonia. Killer T-cells which specifically attack CMV can be separated from the person donating the bone marrow and large numbers can be grown in culture. In one recent study, such antigen-specific killer T-cells grown *in vitro* were used to inject immuno-suppressed patients. Preliminary results indicated that none of the patients suffered from CMV infection after this treatment. One future possibility of this kind of treatment will be to use specific killer T-cells against the HIV virus in order to reduce or prevent the symptoms of AIDS.

Gene Targeting

As our understanding of certain genetic disorders increases, the logical direction of these transplantation techniques will be to correct faulty genes before reintroducing the tissue into the patient. Advances have been made in introducing foreign genes by insertional mutagenesis into cultured skin and blood vessel cells.

Amniocentesis, Infertility and Embryo Transplantation

Genetic abnormalities of foetuses may be identified by culturing cells collected from the amnion during early pregnancy. Some of the amniotic fluid is removed by a process called amniocentesis and the cells are cultured to provide enough material for chromosome analysis. The techniques of animal tissue culture are also directly relevant to *in vitro* fertilization and embryo transplantation that are employed to circumvent some of the problems of infertility.

Cytotoxicity Testing

Tissue culture has been used to screen many anti-cancer drugs since the demonstration in 1950 of clear correlations between the *in vitro* and *in vivo* activities of potential chemotherapeutic agents. Explants of human tumour tissues grown *in vitro* can be tested for chemosensitivity in order to tailor the patient's chemotherapy to suit the individual patient and tumour.

At present, evaluation of the effects of carcinogens, toxins and drugs on specific cell types using *in vitro* models supplements the findings found using animal models. The most extensive use of experimental animals is in the safety evaluation of drugs, pesticides, food additives, industrial chemicals and cosmetics which all have to be screened for safety. Safety evaluation of a single drug or other chemical to the stage at which it is marketed may involve the use of as many as 1000 animals. In the UK, several million experimental animals are likely to be used for the purposes of safety evaluation of chemicals even though tissue culture tests would be more suitable both in terms of cost and for ethical reasons. In addition, there is a scientific justification for the tendency towards cytotoxicity testing in tissue culture, namely the realization that animal models are in many ways inadequate for predicting the effects of chemicals on humans since there are many metabolic differences between species. Cytotoxicity studies involve the analysis of morphological damage or inhibition of zone of outgrowth induced by the chemical being tested.

From the range of applications of animal tissue cultures we have described above, you should be impressed by the variety and importance of these applications. Clearly we can identify the potential of using such systems to provide better health care to both humans and animals. But we can also add to this list the use of these techniques in the *in vitro* genetic manipulation of animals to produce animals with desired characteristics or for cloning desired strains of domestic animals. Thus these techniques are important in agriculture as well as in health care.

Terminology

It is important to clarify some of the terms which are used to describe different methods of *in vitro* cultivation of animals cells.

Tissue, Organ and Cell Culture

Tissue culture is used as a generic term to include the *in vitro* cultivation of organs, tissues and cells. As such, the term is not limited to animal cells, but includes the *in vitro* cultivation of plant cells. '*In vitro* Cultivation of Plant Cells', but will not be mentioned further in this text. Tissue culture can be subdivided into two major categories; organ culture and cell culture.

Organ culture refers to a three-dimensional culture of tissue retaining some or all of the histological features of the tissue *in vivo*. The whole organ or part of the organ is maintained in a way that allows differentiation and preservation of architecture, usually by culturing the tissue at the liquid-gas interface on a grid or gel. Organ cultures cannot be propagated and experiments using organ culture generally involve a large degree of experimental variation between replicates, making it difficult to use organ culture for quantitative determinations.

Cell culture refers to cultures derived from dispersed cells taken from the original tissue. These cultures have lost their histotypic architecture and often some of the biochemical properties associated with it. However, they can be propagated and hence expanded and divided to give rise to replicate cultures. Cell cultures can be characterized and a defined population can be preserved by freezing.

In multicellular organisms cells do not function in isolation, but each cell serves the needs of the whole organism. Hence communication is very important for multicellular existence. *In vivo*, cells in close contact interact with each other, and these interactions are essential in the functioning of tissues and organs. Cells are also in contact with a complex network of secreted proteins and carbohydrates called the extracellular matrix, that fills the spaces between cells. The extracellular matrix helps to bind cells together, provides a lattice through which cells can move, and affects cell behaviour directly. In addition, cells *in vivo* are in contact with hormones and hormone-like factors. This communication between different cell types is lost when cells are cultivated *in vitro*. When cells are grown in culture they are severed from all of the above interactions and can, therefore, not be expected to behave in exactly the way they would *in vivo*.

Adult and Embryonic Tissue

Cultures can be derived from adult tissue or from embryonic tissue. Cultures derived from embryonic tissue generally survive and grow better than those taken from adult tissue. Tissues from almost all parts of the embryo are easy to culture, whereas tissues from adults are often difficult or impossible to culture. Widely used embryonic cells include mouse embryo fibroblasts 3T3 cell lines and the human foetal lung fibroblast cell lines such as MRC-5 (do not worry about the lettering of these cell lines at the moment. You will learn what some of the important abbreviations mean later). Despite the practical advantages of embryonic cell culture, it is important to remember that these cells may not behave in the same way as adult cells, and they need to be characterized extensively.

Embryonic Stem Cells

A more recent development has been the removal of *embryonic stem cells* (ES-cells) from the embryo during the blastocyst stage of development. These cells can be grown in culture for many generations and are of particular interest because they can be manipulated in culture and then re-introduced into embryos.

Adherent or Suspension Cultures

Cells may grow as an adherent monolayer or in suspension. Adherent cells are said to be anchorage-dependent and attachment to a substratum is a prerequisite for proliferation. They are generally subject to contact inhibition, which means they grow as an adherent monolayer and stop dividing when they reach such a density that they touch each other. Most cells, with the exception of mature haemopoietic (haematopoietic) cells and transformed cells, grow in this way.

In contrast to anchorage-dependent cells, cells cultured from blood, spleen or bone marrow adhere poorly if at all to the culture dish. In the body, these cells are held in suspension or are only loosely adherent. It is important to realize this if you are working with this category of cells, since the methods used to propagate these cells are very different to those for adherent cells.

Suspension cultures are easier to propagate, since subculture only requires dilution with medium. Cultures in which cells grow attached to each other or to a substratum have to be treated by a protease to break the bond between cells and substratum. The most commonly used enzyme is typsin. Clearly, freely suspended cultures do not require trypsinization. They are, therefore, also easier to harvest.

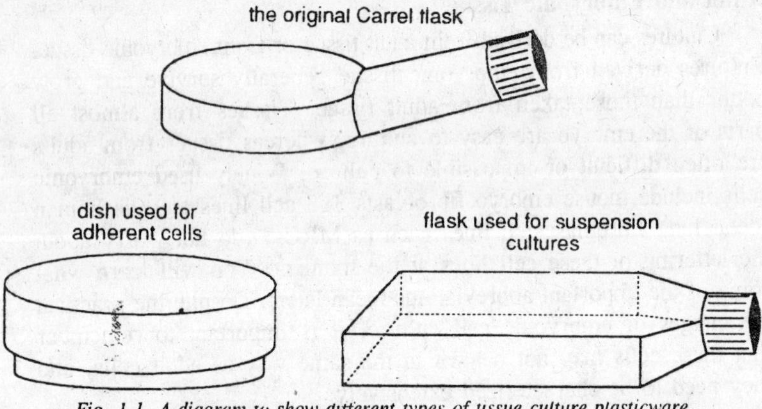

Fig. 1.1. A diagram to show different types of tissue culture plasticware.

What we hoped you would identify is the fact that all the vessels shown have a large surface area to which cells may adhere. This is, of course, vital for anchorage-dependent cells if we are to achieve a reasonable cell yield. The main difference is that of the access into the vessel for inoculation and harvesting. Clearly it is easier to inoculate and harvest culture from the dish than from the two flasks shown because by removing the lid we can get to the whole substratum which bears the cells. There is, however, a greater chance of the culture becoming contaminated if it is exposed in this way.

Tumour or normal tissue

Cells derived from normal tissue may not grow indefinitely *in vitro*. This will be covered in more detail in the next section. The only human cells which will grow indefinitely in culture are those obtained from tumour tissue. This means that work with non-malignant human cells needs a constant source of fresh tissue and the problem is that normal human tissue is relatively difficult to obtain. The restriction with respect to the source of tissue means that in practice only a few cell types are commonly used in human tissue culture. Cells which will grow indefinitely *in vitro* are said to be 'immortal'.

STAGES IN CELL CULTURE

Primary Cell Culture

Within the category of cell culture, we can make further subdivisions. We will discuss these in more detail now. The first step in preparing any culture is the sterile dissection of the tissue from the organism concerned. The tissue is chopped into pieces of around 1 mm^3 which are put on a dish. At this stage the tissue is called an *explant*, which simply means that it is a tissue taken from its original site and transferred to an artificial medium for growth. The cells of the tissue can be isolated and disaggregated by mechanical, chemical or enzymatic digestion of animal tissue. When these cells are induced to grow *in vitro*, a primary cell culture results and these cells are generally still fairly representative of the original tissue. The culture is called a primary culture until it is subcultured for the first time, after which it becomes a secondary culture.

Two important disadvantages of primary cell culture over some of the other types of cells culture are that:

1. A mixture of cell types is generally present (since most tissues contain mixtures of cells), in other words it is heterogeneous;
2. Primary cell culture requires the recurrent sacrifice of animals.

Passaging Cells

Adherent cells grow as a monolayer until they reach confluence. The kinetics of animal cell growth *in vitro* follows a similar pattern to the classical kinetics demonstrated in cultures of bacteria. When cells are taken (from a tissue, primary culture or stationary phase culture) there is at first a *lag phase* of some hours or days before the

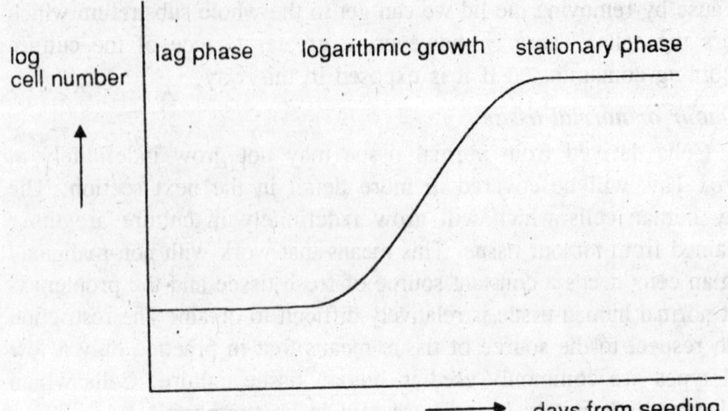

Fig. 1.2. Diagrammatic representation of passaging cells.

cells begin to grow. Growth then proceeds steadily with the population doubling every 15-20 hours in the case of fast-growing cells. This phase of exponential growth is known as the *logarithmic phase*. At the end of this phase the maximum population is reached and cells enter the *stationary phase* or plateau phase in which virtually no growth occurs. At this stage the cells can be trypsinized and reintroduced into fresh media. This is called 'passaging' or 'sub-culturing'. Ideally cells should be passaged during the late log phase of cell growth.

The *passage number* is the number of times this procedure is performed after the original isolation of cells from the primary source. When a primary culture reaches confluence, the cells should be trypsinized and re-seeded in a fresh dish. At this particular stage it is called a secondary culture and more generally from now on through subsequent passaging it is called a *cell line*. A cell line arises from a primary culture at the time of the first successful subculture and the term implies that cultures consist of lineages of cells originally present in the primary culture. Since the primary culture was heterogeneous, the cell line arising from the primary culture can also be heterogeneous. Selection or cloning of cells with particular properties or markers can be performed on either the primary culture or the cell line, resulting in different *cell strains*. Thus the term cell strain implies some sort of selection or cloning has taken place.

The Fate of Primary Cultures

Some cells are capable of an unlimited number of cell divisions *in vitro* as long as they are supplied with nutrients. We can, therefore, sub-culture such cells indefinitely and there is no limit to the number of passages. These are described as continuous or immortalized cell lines or strains. Other cells are only capable of limited numbers of cell divisions after which the culture stops dividing. These are called finite cell lines or strains. The terms finite or continuous are used as prefixes if the status of the culture is known. Immortalized cell lines were previously sometimes called *established cell lines*. Actually, the term continuous cell lines is somewhat confusing because the term "continuous" in bioprocess technology is usually used to describe the cultivation of cells in a "continuous" rather than a batch system. We would, therefore, prefer to use the term immortalized cell lines to describe cell lines that can be sub-cultured (passaged) indefinitely.

The replicative capacity of cultured cells varies enormously depending on the cell type and the species. Many cells can be passaged a large number of times, whereas others will die at an early stage. There are some interesting differences between species in their reaction to culturing. In rodent tissue culture, it is not uncommon for cells to divide indefinitely whereas cells of normal human tissue never give rise to continuous cell lines and only tumour cells grow indefinitely. Chicken cells are difficult to maintain in culture for any length of time and die after only a few population doublings. The cause for this limited growth potential is not understood, and no nutritional regime has yet been discovered to solve the problem.

Human cell culture

In 1961, Hayflick and Moorhead studied the potential of human foetal lung fibroblasts to divide in culture by counting the cells at each passage starting with the explant from human tissue. They found a slow increase in the growth rate (phase I). During this phase, some cells die, and other cells grow. If the culture is continually diluted they grow at constant rate for an average of 50 generations (phase II) after which the growth rate begins to slow down. The ensuing period of increasing cell death (phase III) ultimately leads to complete death of the culture. Thus, human cells could be 'passaged' or 'sub-cultured' for about 50 population doublings before growth stopped and senescence occurred. Since then, other human cell types have been studied, and the results have generally been similar, although the exact number of doublings depends on the type of cell, its stage of differentiation and origin. The number of population doublings for human cells is generally between 20 and 80 but it can be much shorter. At the end of this stage, the cells first start to look strange and a few weeks later the culture always dies even when apparently provided with all the right nutrients. The limited replicative capacity of human cells in culture is sometimes called the *Hayflick effect*, after its discoverer.

Further experiments indicated that the loss of replicative capacity has an interesting link with the process of aging. The average number of population doublings achieved was 50 for cells derived from human foetal lung fibroblasts, but only 20 for cells derived from adult lung fibroblasts. Moreover, experiments with human skin fibroblasts, in 1970, indicated that the number of doublings achieved in culture is a function of the donor's age. A linear decrease in the life span of the cells was found with increasing age of donor. These experiments indicated that there does appear to be some relationship between the aging of cells in culture and the aging of cells in the human body and suggest that cells may have a built in "time clock" which tells them how long to survive.

You would have $10^5 \times 2^{50}$ (approximately 10^{20}) cells, thus even though the lifetime of these cells is limited, the culture can still be studied for a long time, and may be a valuable research tool. However, such cultures still have a number of disadvantages. As the cultures gets older, there are changes in the behaviour of the cells so controlled studies are difficult. The finite lineage of cells originating from such primary cultures have different properties at different stages of passaging, so it is important to carefully record the passage number. Also, the limited life-span of these cells means that there is still a need to find fresh tissue when the cells die. The ideal cell culture system would grow indefinitely without variation in cell phenotype.

Rodent cell culture

A fraction of the cell population from a primary culture derived from a rodent may continue to grow beyond the point when cells from many other sources would have stopped growing and died. The culture is designated as continuous or immortal when it has been sub-cultured at least 70 times at an interval of three days between each sub-culture. This often happens in rodent tissue culture experiments and cultures of rodent embryo cells routinely give rise to immortal cultures.

When a culture is prepared from rodent tissue, there is some initial cell death coupled with the emergence of healthy growing cells. As these are diluted and allowed to continue growth, they soon begin to lose growth potential and most of the cells die. The culture at this stage is said to undergo crisis. However, occasionally a variant population of cells emerges from this phase and these cells continue

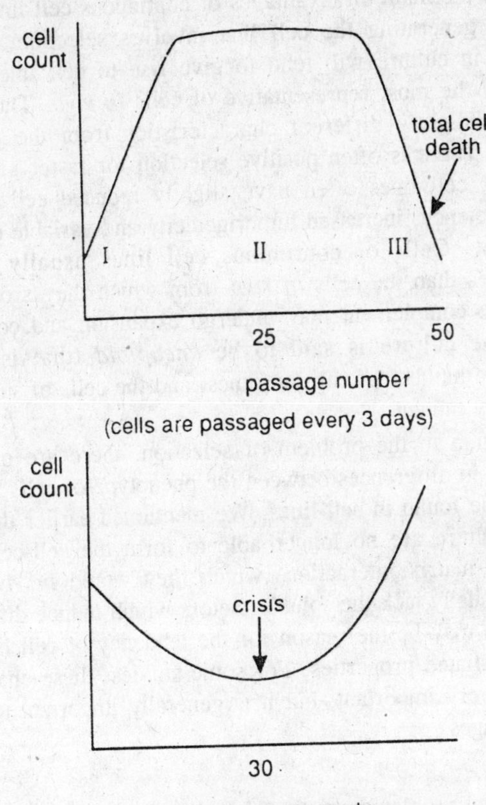

Fig. 1.3. Phages in the establishment of human and rodent cell cultures.

to grow until their progeny overgrow the culture. These cells constitute an immortalized cell line which will grow forever if appropriately diluted and fed with nutrients. A protocol which is much used to establish rodent cell lines is one whereby 3×10^5 cells are transferred every three days to new dishes. The lines made by this protocol are called 3T3 lines.

There is a smooth transition to continuous culture, although the origin of the cells which continue to grow at this stage is not entirely clear. They may result from a mutational event, or alternatively it may be that immortalized cells were already present in the original culture, but that they were masked by the presence of finite populations of cells.

One of the main disadvantages of continuous cell lines is that the process of generating the cell line involves selection. Cells which grow well in culture will tend to give rise to cell lines, and these may not be the most representative of cells *in vivo*. Thus, cell lines often have slightly different characteristics from the original cell population. There is often positive selection for faster growing cells. In addition, cell lines often have slightly reduced cell size, higher cloning efficiency, increased tumorigenicity and variable chromosome complement. Cells of continuous cell lines usually have more chromosomes than the cells *in vivo* from which they arose and their chromosome complement may undergo expansion and contraction in culture. The culture is said to be *aneuploid* (that is having an inappropriate number of chromosomes) and the cells of such a culture are obviously mutants.

In addition to the problem of selection, there are other factors which result in differences between the phenotype of cells *in vivo*, and the phenotype found in cell lines. We mentioned earlier that cells are grown in culture are no longer able to form the cell-cell and cell-extracellular matrix interactions which they would *in vivo* and that these cells usually lack the soluble factors which induce differentiation *in vivo*. This may be the reason for the tendency of cell lines to lose their differentiated properties. For some studies, these characteristics may not be very important, but it is generally important to be aware of these changes.

Cell Cloning

One of the problems mentioned earlier for primary cell culture is that these cultures do not contain only one cell type, but often include a number of different cell types while many biological studies require populations of a single cell type. A number of methods exist to separate different cell types. The traditional approach is to isolate a pure cell strain from cells in continuous culture by a process called *cloning*. During cloning, a population of cells is derived from a single cell by mitosis to produce a genetically homogeneous clone which can then be characterized and stored. The uniformity of the cells within the cell clone and the potential to increase cell number opens up a wide range of experimental possibilities. A number of different cloning protocols exist based on either physical separation or the use of selective conditions. In *dilution cloning*, the cell are seeded at low density and are incubated until colonies form. They are then isolated and propagated into cell strains.

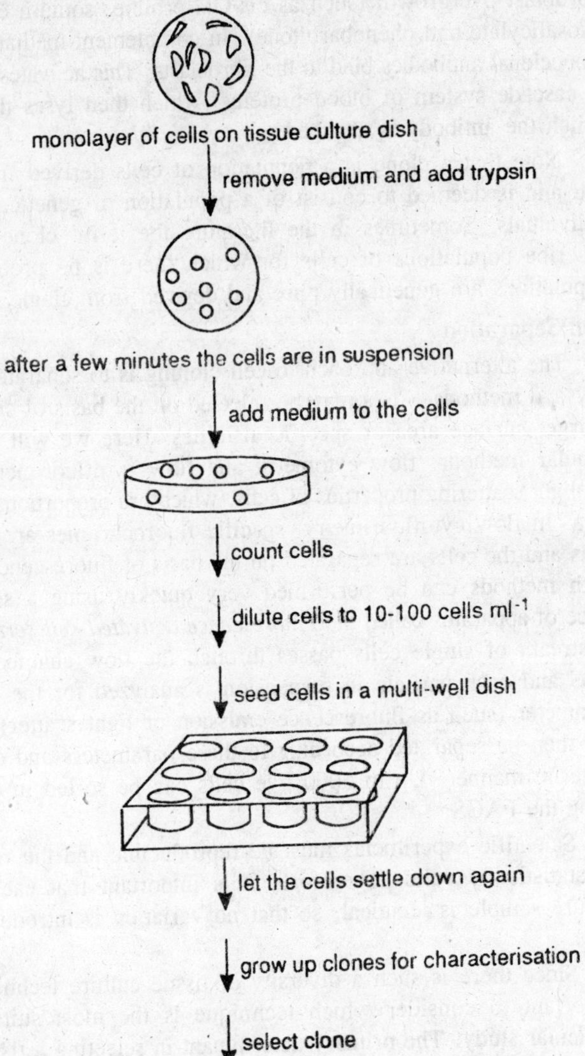

Fig. 1.4. The technique of dilution cloning. The process is designed to deposite a single cell in individual wells in the multi-well dish.

Another way of isolating cell clones is to use *selective media*. One of the main problems is that for instance fibroblasts tend to grow well in culture and therefore contaminate other cell types. Fibroblasts can be eliminated either by complement-mediated lysis using monoclonal antibodies against fibroblasts or by using chemicals which suppress fibroblast overgrowth (such as cis-OH-proline, sodium ethylmercurithiosalicylate and phenobarbitone). In complement-mediated lysis, the monoclonal antibodies bind to the fibroblasts. This activates complement (a cascade system of blood proteins) which then lyses those cells to which the antibody is attached.

Note that a clone is a population of cells derived from a single cell and is deemed to consist of a population of genetically identical individuals. Sometimes in the literature the term' clone' is used to describe populations of cells for which there is no proof that these populations are genetically pure and derived from single cells.

Cell Separation

The alternative approach to cell cloning is to separate cells using physical methods. Cells may be selected on the basis of size, density, charge, surface area or specific affinities. Here we will discuss two popular methods; flow cytometry and flow cytofluorimetry measure the light scattering properties of cells, which are proportional to surface area. In flow cytofluorimetry, specific fluorochromes are attached to cells and the cells are separated on the basis of fluorescence emission. Both methods can be performed very quickly using a sophisticated piece of apparatus called the *Fluorescence activated cell sorter* (FACS). A stream of single cells passes through the flow chamber of the F ACS and each particle in suspension is analyzed for the appropriate parameter (such as fluorescence emission or light scatter). The cells can then be separated according to these parameters and collected in a sterile manner. Up to 5000 live cells can be sorted in one second using the FACS.

Scientific experiments must be reproducible and the results must be statistically valid. Therefore, it is important that each replicate culture sample is identical, so that no variation is introduced at this level.

Since there is such a diversity of tissue culture techniques, it is important to consider which technique is the most suitable for a particular study. The primary determinant in selecting a tissue or cell line for study is the nature of the observations to be carried out. General cellular processes, such as analysis of DNA synthesis, can be carried out on almost any cell type whereas more specialized processes can only be carried out on the appropriate cell type, for example analysis of the control of milk production is only feasible using mammary epithelial cells.

Animal Tissue Culture in Perspective

In vitro culture of animal cells and organs enables experiments to be conducted at greater uniformity than can be achieved using whole animals, leading to improved reproducibility between successive experiments. In addition, tissue culture avoids many of the moral and ethical questions of animals experimentation and allows experiments on human tissue which would not be possible *in vivo*. However, it is important to remember that tissue culture models still involve the killing of animals to supply material.

The *in vitro* cultivation of animal cells also has a number of disadvantages over *in vivo* models. Firstly, tissue culture requires a certain amount of skill - culture techniques must be carried out under strict aseptic conditions because animal cells grow less rapidly than many of the common contaminants such as bacteria and yeast. Secondly, the cost of animal tissue culture reagents, and the labour involved in maintaining animal cells in culture, are often greater than the costs involved in most animal work. (This is certainly the case for experiments on rodents, but *in vivo* experiments using dogs, cats or primates are a lot more costly than tissue culture work). Thirdly, cell lines may not reflect the situation *in vivo*, and whatever system is being studied, great care must be taken to analyze the significance of any *in vitro* findings. It is not possible to reproduce the conditions in the living animal exactly. Finally, tissue culture is not always possible, since suitable cell lines may not be available.

Overall, tissue culture and *in vivo* studies complement each other and both have their advantages and disadvantages in specific situations. It is important to consider which technique is most suitable to address any specific question and to use the most appropriate model for the situation. Tissue culture cells do not behave in exactly the same way as cells do *in vivo* but so long as the limitations of the model are appreciated, tissue culture is still a very valuable tool.

2

EQUIPMENTS

The specific needs of a tissue culture laboratory, like most labs, can be divided into three categories: (1) essential—you cannot perform a job without them; (2) beneficial—the work would be done better, more efficiently, quicker, or with less labor; and (3) useful—it would make life easier, improve working conditions, reduce fatigue, enable more sophisticated analyses to be made, or generally make your working environment more attractive. Equipment that might be used in tissue culture is listed in Table 1 in the grouping suggested above. Remember two main points: assuming you need it and can afford it, you must be able to get it into the room (*access*) and you must have space for it (*accommodation*). Suppliers of items of equipment will be found in the "Trade Index—Source of Material" at the end of this boll.

ESSENTIAL EQUIPMENT

Incubator

This should be large enough, probably 200 1 (6 ft^3) per person, have forced air circulation, temperature control ± 0.5°C, and a safety thermostat which cuts out if the incubator overheats or, better, which regulates it if the first fails. It should be corrosion resistant, e.g., stainless steel (anodized aluminum is acceptable for a dry incubator), and easily cleaned. A double cabinet, one above the other, independently regulated, gives you more accommodation with the added protection that if one-half fails the other can still be used. This is also useful when you need to clean out one compartment.

Incubation Temperature

The optimal temperature for cell culture is dependent on (1) the body temperature of the animal from which the cells were obtained;

(2) any regional variation in temperature (e.g. skin may be lower); and (3) the incorporation of a safety factor to allow for minor errors in incubator regulation. Thus, the temperature recommended for most human and warm-blooded animal cell lines is 3.6°C, close to body heat but set a little lower for safety.

Avian cells, because of the higher body temperature in birds, should be maintained at 38.5°C for maximum growth but will grow quite satisfactorily, if more slowly, at 36.5°C.

Cultured cells will tolerate considerable drops in temperature can survive several days at 4°C and can be frozen and cooled to −196°C, but they cannot tolerate more than about 2°C above normal (39°C) for more than a few hours, and will die quite rapidly at 40°C and over.

Epidermal cells from the mouse may grow better at a slightly lower temperature, etc., 33°C.

In general the cells of poikilothermic animals have a wide temperature tolerance but should be maintained at a constant level within the normal range of the donor species. This requires incubators with cooling as well as heating as the incubator temperature may need to be below ambient (e.g. for fish). Cooling capacity should be sufficient to lower the temperature about 2°C, or more, below ambient so that regulation is performed by the heater circuit, which is more sensitive.

If necessary, poikilothermic animal cells can be maintained at room temperature, but the variability of the ambient temperature in laboratories makes this undesirable.

Regulation of temperature should be kept within ± 0.5°C; consistency is more important than accuracy. Cells will grow quite well between 33° and 39°C but with naturally vary in growth rate and metabolism. The incubation temperature should be kept constant both in time and at different parts of the incubator. Water baths gives the most accurate control of temperature, but temperature control. They are, therefore, seldom used and incubators are preferable. The air should be circulated by a fan to give even temperature distribution, and cultures should be placed on perforated shelves and not on the floor or touching the sides of the incubator.

Sterilizer

The simplest *sterilizer* is a domestic pressure cooker which will generate 1 atm (15 lb/in^2) above ambient. More complex autoclaves exist, but the main consideration is the capacity; will it accommodate all you want to do? A simple bench top autoclave may be sufficient but a larger model with a timer and a choice of presterilization and poststerilization evacuation will give more capacity and greater flexibility in use. A "wet" cycle (water, salt solutions, etc.) is performed without evacuation before or after sterilization. Dry items (instruments, swabs, screw caps, etc.) require the chamber to be evacuated before sterilization, to allow efficient access of hot steam, and should be evacuated after sterilization to remove steam and promote subsequent drying; otherwise the articles will emerge wet, leaving a trace of contamination from the condensate on drying. To minimize this risk always use deionized or reverse osmosis water to supply the autoclave.

If you require a high sterilization capacity (300, 1, 9 ft^3, or more), buy two smaller autoclaves rather than one large one, so that during routine maintenance and accidental breakdowns you still have one functioning machine. Furthermore, a smaller machine will heat up and cool more quickly and can be used more economically for small loads. Leave sufficient space around them for maintenance and ventilation and provide adequate air extraction to remove heat and steam.

Refrigeration and Freezers

Usually a domestic item will be found to be quite efficient and cheaper than special laboratory equipment. Domestic refrigerators are available with no ice box ("larger refrigerators") giving more space and eliminating the need for defrosting. However, if you require a lot of accommodation (400, 1 12 ft^3, or more), a large hospital (blood bank) or *catering freezer* may be better. While *autodefrost freezers* may be bad for some reagents (enzymes, antibiotics, etc.) they are very useful for most tissue culture stocks where their bulk and nature precludes severe cryogenic damage. Conceivably, serum could deteriorate during oscillations in the temperature of an autodefrost freezer, but in practice it does not seem to. Many of the essential constituents of serum are small proteins, polypeptides, and simpler organic and inorganic compounds which may be insensitive to cryogenic damage.

Microscope

It cannot be ovestressed that in, spite of considerable and highly desirable progress toward quantitative analysis of cultured cells, it is still vital to look at them regularly. A morphological change in often the first sign of deterioration in culture and the characteristic pattern of microbiological infection is easily recognized.

A simple *inverted microscope* is essential. Make certain that the stage is large enough to accommodate large roller bottles in case you should require them. There are many simple and inexpensive inverted microscopes in the market; but if you foresee the need for photography of living cultures, then you should invest in one with high-quality optics, a long working distance phase-contrast condenser and objectives, with provision to take a camera.

Washing-up Equipment

Soaking baths or sinks

Soaking baths or *sinks* should be deep enough so that all your glassware (except pipettes and large aspirators) can be totally immersed in detergent during soaking, but not so deep that the weight of glass is sufficient to break smaller items at the bottom, e.g., 400 mm (15 in) wide × 600 mm (24 in) long × 300 mm (12 in) deep.

If you are designing a lab from scratch, then you can get sinks built in of the size that you want. Stainless steel or polypropylene are best, the former if you plan to use *radioisotopes* and the latter for hypochlorite disinfectants.

Washing sinks should be deep enough (450 mm, 18 in) to allow manual washing and rinsing of your largest items without having to stop too far to reach into them, and about 900 mm (3 ft) From the floor to rim. It is better to be too high than too low; a short person can always stand on a raised step to reach a high sink but a tall person will always have to bend down if the sink is too low. There should be a raised edge around the top of the sink to contain spillage

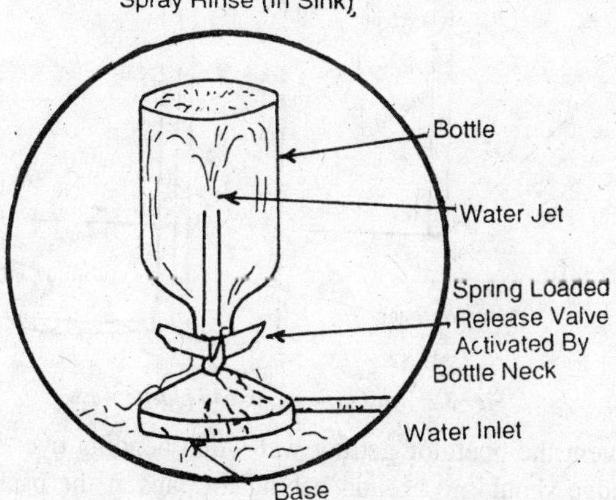

Fig. 2.1. Figure showing spray rinse (in sink).

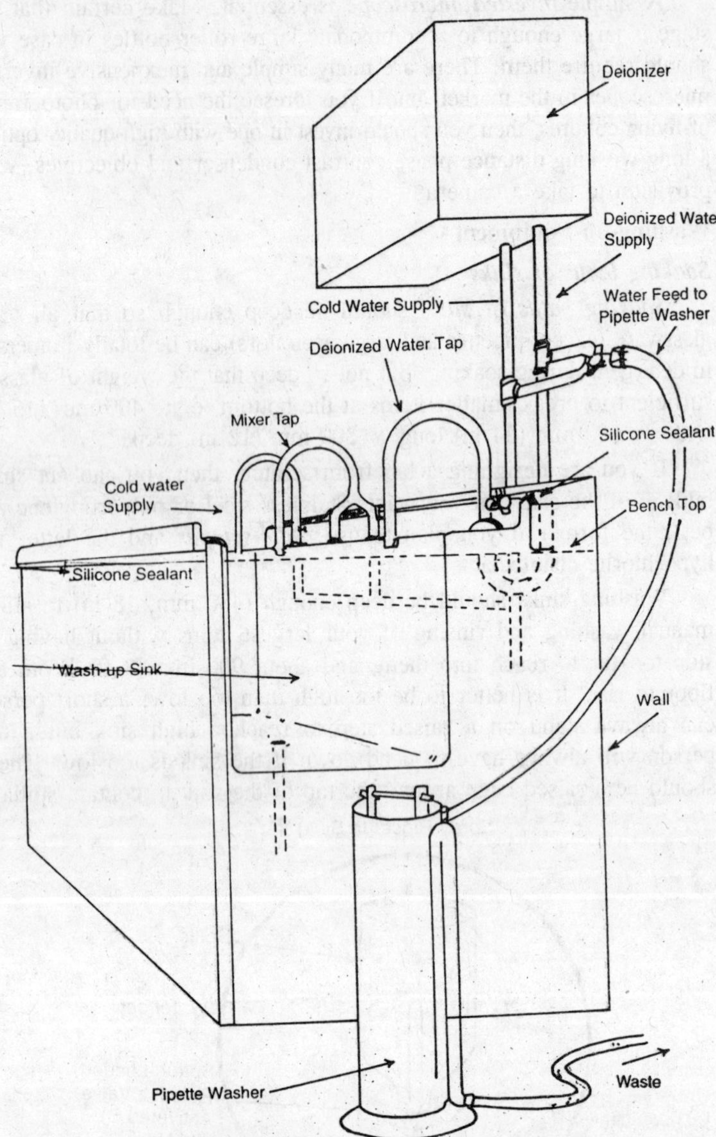

Fig. 2.2. Washing-up sink and pipette washer.

and prevent the operator getting wet when bending over the sink. The raised edge should go around behind the taps at the back.

Each washing sink will require four taps; a single cold water, combined hot/cold mixer, a cold hose connection for a rinsing device, and a nonmetallic tap for deionized water, from a reservoir above the sink. A centralized supply for deionized water should be avoided as the pipe work can build up dirt and algae and is difficult to clean.

Pipette cylinders

These should be made from polypropylene and be freestanding, distributed around the lab, one per work station.

Pipette washer

Following an overnight soak in detergent, reusable pipettes are easily washed in a standard siphon-type washer. This should be placed at floor rather than bench level to avoid awkward lifting of the pipettes and connected to the deionized water supply so that the final rinse can be done in deionized water. If possible a simple changeover valve should be incorporated into the deionized water feed line.

Pipette drier

If a stainless steel basket is used in the washer, this may then be transferred directly to an electric drier. Alternatively, pipettes can be dried on a rack or in a regular drying oven.

Sterilizing and Drying Oven

Although all sterilizing can be done in an autoclave, it is preferable to sterilize pipettes and other glassware by dry heat, avoiding the possibility of chemical contamination from steam condensate or corrosion of pipette cans. This will require a high-temperature (160–180°C) fan-powered oven to ensure even heating throughout the load. As with autoclaves, do not get an oven that is too big for the size of glassware that you use. It is better to use two small ovens than one big one; heating is easier, more uniform, quicker, and more economical when only a little glassware is being used. You are also better protected during breakdowns.

Water Purification

Pure water is required for rinsing glassware, dissolving powdered media, or diluting concentrates. The first of these is usually satisfied by deionized water but the second and third require a higher degree of purity, demanding a three-or four-stage process. The important principle is that each stage is qualitatively different; reverse osmosis may be followed by charcoal filtration, deionization, and micropore filtration (e.g. *Millipore*) or *distillation* (with a silica sheathed cement) may be

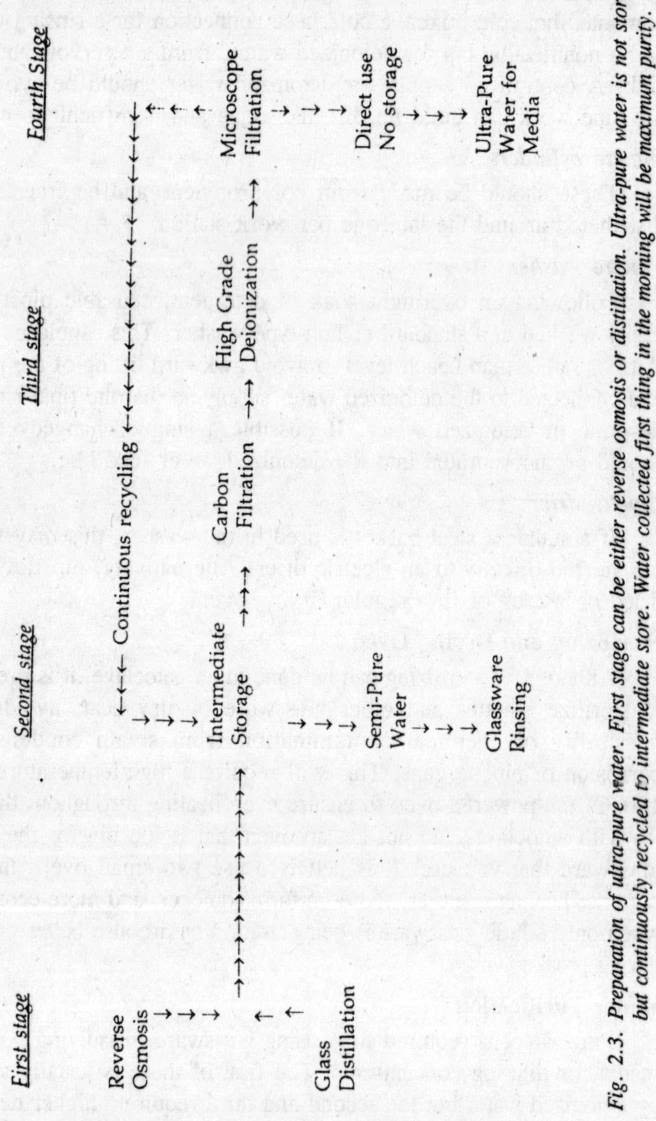

Fig. 2.3. *Preparation of ultra-pure water. First stage can be either reverse osmosis or distillation. Ultra-pure water is not stored but continuously recycled to intermediate store. Water collected first thing in the morning will be maximum purity.*

substituted for the first stage. *Reverse osmosis* is cheaper if you pay the fuel bills but if not, distillation is better and more likely to give a sterile product. If reverse osmosis is used, the type of cartridge should be chosen to suit the pH of the water supply.

The deionizer should have conductivity meter monitoring the effluent to indicate when the cartridge must be changed. Other cartridges should be dated and replaced according to the manufacturer's instructions.

Purified water should not be stored but recycled through the apparatus continually to preclude algal infection. Any tubing or reservoirs in the system should be checked regularly (every 3 months or so) for algal infection, cleaned out with hypochlorite and detergent (e.g.., Chloros), and thoroughly rinsed in purified water before reuse.

Water is the simplest but probably the most critical constituent of all media and reagents, particularly with serum-free media.

Centrifuge

Periodically, cell suspensions require centrifugation to increase the concentration of cells or to wash off a reagent. A small bench-top centrifuge, preferably with proportionally controlled braking, is sufficient for most purposes. Cell sediment satisfactorily at 80-100 g; higher g may cause damage and promote agglutination of pellet. A large-capacity refrigerated centrifuge, 4 × 1 liters or 6 × 1 liters, will be required, if large scale suspension cultures are contemplated.

Cell Freezing

The procedures for *cell freezing* will be dealt with in detail elsewhere, but the basic facilities should be considered here. The freezing process can be carried out satisfactorily without sophisticated equipment, but storage requires a properly constructed liquid N_2 freezer and storage dewar. Freezers range in size from around 25 1 to 500 1, i.e. 250-15000 1-ml ampules. It is best to freeze a minimum of five ampules for each cell strain, 20 for a commonly used strain, and 100 for one in continuous use. A capacity of 1,200-1,500 is appropriate for most small laboratories.

The choice of freezer is determined by three factors: (1) capacity (no. of ampules); (2) economy and static holding time (the time taken for all the liquid N_2 to evaporate)—both of which are governed by the evaporation rate, which in turn is dependent on the frequency of access; and (3) convenience of access. Generally speaking, 2 is inversely proportional to 1 and 3. There are two main types of freezer: narrow-necked with slow evaporation but with more difficult access, and wide-necked with easier access but three times the evaporation rate. If the cost of liquid nitrogen and its supply presents to problem, then a wide-necked freezer may be more convenient although the holding time will only be about 1 wk to 10 d. A narrow-necked freezer, on the other hand, will be more economical and last up to 2 months if N_2 supplies run out.

If you require bulk-storage (~ 10,000 ampules), then you will need to consider a vessel of around 300 1 capacity. Wide-necked freezers are most common in this size because of the mechanical difficulties in operating narrow-necked freezers of high capacity, but the latter are available and will save a considerable amount in expenditure on liquid N2. At 300 1 the evaporation rate is approximately 10 1/d in a wide-necked freezer.

The advantages of gas-phase and liquid-phase storage will be discussed elsewhere, but one major implication of storing in the *gas phase* is that the liquid phase is necessarily reduced to the space below your ampule storage area, usually 20-30% of the full volume. Hence, the static holding time is reduced to one-third or one-fifth of that of the filled freezer, filling must be carried out more regularly, and the chances of accidental thawing are increased. Where the investment is higher (many ampules or rare cell strains) automatic alarm systems should be fitted and, for the high-capacity freezers, an automatic filling system is recommended. However, automotive systems can fail and a twice-weekly check of liquid levels with a dipstick should be maintained and a record kept.

An appropriate storage vessel should also be purchased to enable a backup supply of liquid N_2 to be held. The size of this depends: (1) on the size of the freezer; (2) the frequency and reliability of delivery of liquid N_2; and (3) the rate of evaporation. A 40 1, wide-necked freezer will require about 20-30 1/wk, so a 50 1 dewar flask (or two 25 1) flask, which are easier to handle) is advisable. A 35 1, narrow-necked freezer, on the other hand, using 5-10 1 w/k will only require a 25 dewar. Larger freezer are best supplied on line from a dedicated storage tank, e.g., a 160 1 storage vessel linked to a 320 1 freezer with automatic filling and alarm.

Beneficial Equipment

The above describes the essential equipment for modest tissue culture facility; but there are several items of equipment, which, if your budget will stretch to them, will make your laboratory easier to use and more efficient.

Laminar Flow Hood

Usually one hood is sufficient for two to three people. A horizontal flow hood is cheaper and gives best sterile protection to your cultures, but for potentially hazardous materials (radioisotopes, carcinogenic or toxic drugs, virus-producing cultures, or any primate (including human) cell lines), a Class II biohazard cabinet should be used. It is important to consult local biohazard regulations before equipping, as legal requirements and recommendations vary.

Choose a hood that is: (1) large enough (usually 1,200 mm (4 ft) wide × 600 mm (2 ft) deep); (2) quite (noisy hoods are more fatiguing); (3) easily cleaned both inside the working area and below the work surface in the event of spillage; and (4) comfortable to sit at (some cabinets have awkward ducting below the work surface, which leaves no room for your knees, or have screens which obscure your vision). The front screen should be able to be raised, lowered, or removed completely to facilitate cleaning and handling bulky culture apparatus. Remember, however, that a biohazard cabinet will not give you, the operator, the required protection if you remove the front screen.

Insist that you be allowed to examine and sit a hood, as if using it, before committing yourself to purchase. Can you get your knees under it while sitting comfortably and close enough to work? Is there a foot rest in the correct place? Is your head conveniently placed to see what you are doing? Is the work surface perforated and will this give you trouble with spilage (a solid work surface vented at front and back may be better). If the work surfaces are lifted, are the edges sharp or rounded so that you will not cut yourself when cleaning out the hood? Is the lighting convenient and adequate?

Cell Counter

A *cell counter* is a great advantage when more than two or three cell lines are carried and is essential for precise quantitative growth kinetics. Several companies now market models ranging in sophistication from simple particle counting up to automated cell size analysis. For routine counting, the Coulter "D Industrial" is more than adequate and much less expensive than equipment with cell sizing facilities.

Vacuum Pump

A *vacuum pump* or simple tap siphon saves a lot of time and effort when handling large numbers of cultures or large fluid volumes. Tap siphons require a minimum of 6 m (20 ft) head of water to create sufficient suction, but are by far the cheapest, simplest, and more efficient way to dispose of nonhazardous tissue culture effluent. If you do not have sufficient water pressure, or are handling potentially hazardous material, use a vacuum pump similar to that supplied for collected in a reservoir into which a sterilizing agent such as glutaraldehyde or hypochlorite may be added when work is finished and at least 30 min before the reservoir is emptied. A drying agent hydrophobic filter, or second trap placed in the line to the pump prevents fluid being carried over. Do not draw air through a pump from a reservoir containing hypochlorite as the free chlorine will corrode the pump and could be toxic. Avoid vacuum lines; if they become contaminated with fluids, they can be very difficult to clean.

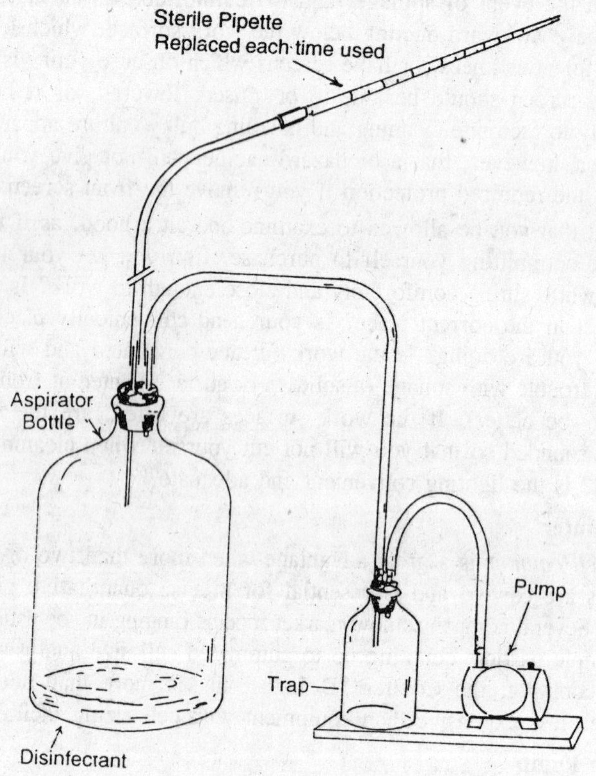

Fig. 2.4. Vacuum pump assembly for withdrawing spent medium, etc.

A peristaltic pump may be used instead of a vacuum pump. No trap is required, effluent can be collected directly into disinfectant, and there is less chance of discharging aerosol into the atmosphere. However, the pump tubing should be checked regularly for wear and the pump operated by a self cancelling foot switch.

Always switch on the pump before inserting a pipette in the tubing to avoid effluent running back.

CO_2 Incubator

Although incubations can be performed in sealed flasks in a regular dry incubator or hot room, some vessels, e.g., petri dishes or multiwell plates, require a controlled atmosphere with high humidity and elevated CO_2 tension. The cheapest way of controlling the gas phase is to place the cultures in a plastic box, a desiccator, or an anaerobic jar.

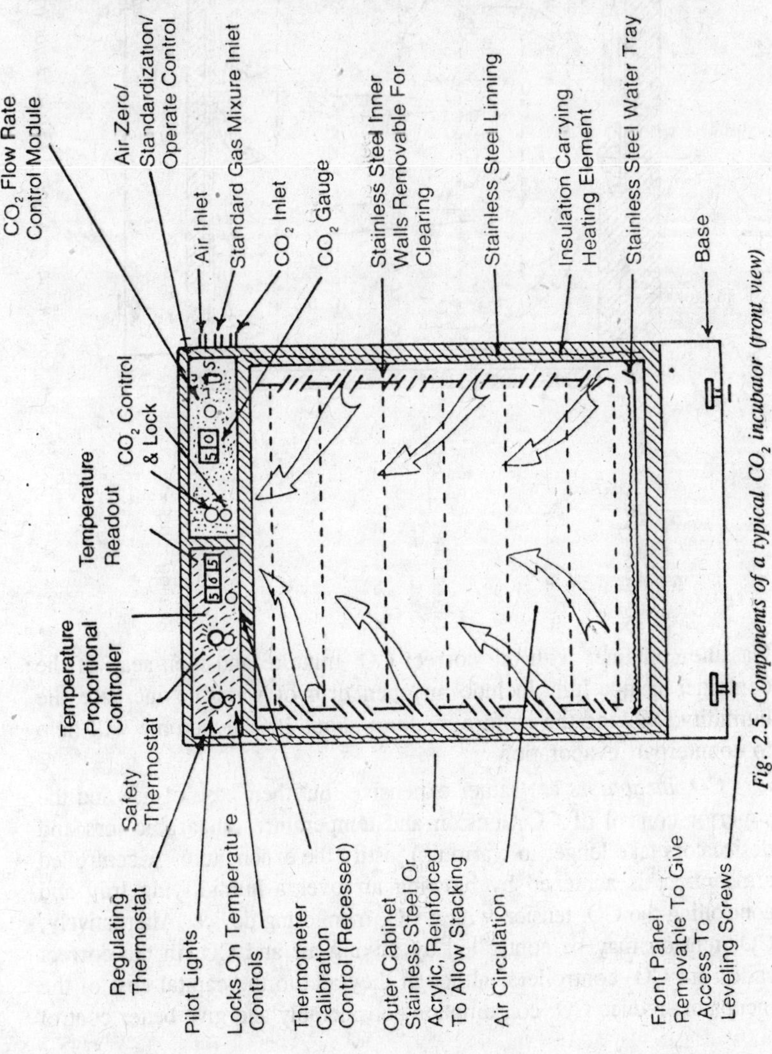

Fig. 2.5. Components of a typical CO_2 incubator (front view)

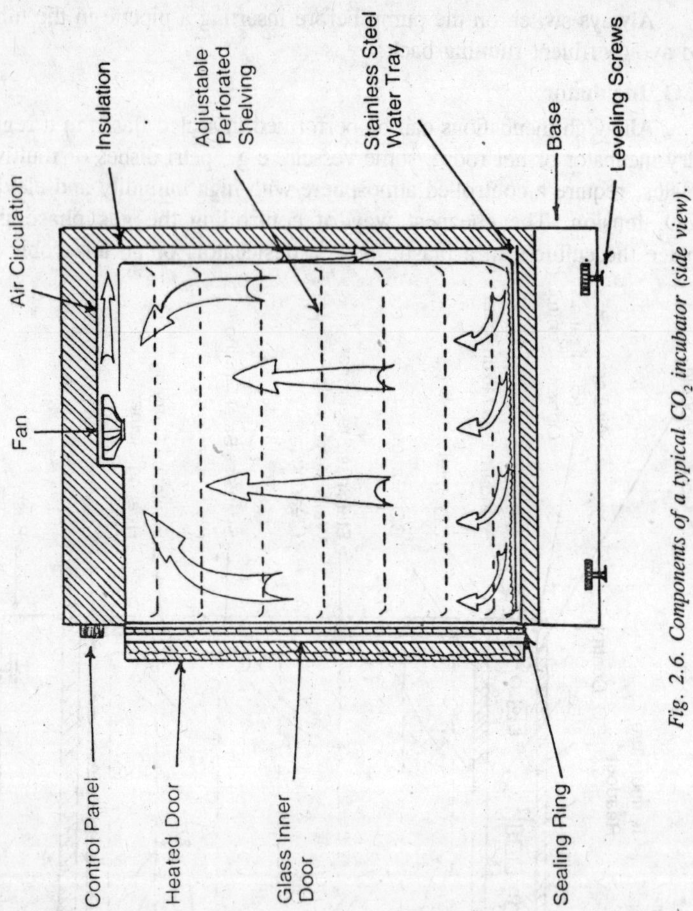

Fig. 2.6. Components of a typical CO_2 incubator (side view).

Gas the container with the correct CO_2 mixture and then seal. If the container is not full, include an open dish of water to increase the humidity. Making the culture medium about 10% hypotonic will help to counterpart evaporation.

CO_2 incubators are rather expensive, but their ease of use and the superior control of CO_2 tension and temperature (anaerobic jars and desiccators take longer to warm up) justify the expenditure. A controlled atmosphere is achieved by blowing air over a humidifying tray and controlling the CO_2 tension with a CO_2 monitoring device. Alternatively, CO_2 tension may be controlled by mixing air and CO_2 in the correct ratio; but CO_2 controllers, although they add to the capital cost of the incubator, reduce CO_2 consumption considerably and give better control

and recovery after opening the incubator. They function by drawing air from the incubator into the sample chamber, determining the concentration of CO_2, and injecting pure CO_2 into the incubator to make up any deficiency. Air is circulated around the incubator to keep both the CO_2 although some manufacturers produce self-calibrating detectors.

Since humid incubators require regular cleaning, the interior should dismantle readily without leaving inaccessible crevices or corners.

Preparation and Quality Control

A course and a fine balance and a simple pH meter are useful additions to the tissue culture area for the preparation of media and special reagents. Although a phenol red indicator is sufficient for monitoring pH in most solutions, a pH meter will be required when phenol red cannot be used, e.g., in preparation of cultures for fluorescence assays.

One of the most important physical properties of culture medium, and one that is often difficult to predict, is osmolality. An osmometer is therefore, a useful accessory to check solutions as they are made up, to adjust new formulations, or to compensate for the addition of reagents to the medium. They usually work by freezing-point depression or elevation of vapour pressure. Choose one with a low sample volume ($\leq$ 1 ml) since on occasion you may want to measure a valuable or scarce reagent

Upright Microscope

An *upright microscope* may be required, in addition to an inverted microscope, for chromosome analysis, mycoplasma detection, and autoradiography. Select a high-grade research microscope, such as the Reichert Polyvar, with regular brightfield optics up to × 100 objective magnification, phase-contrast up to at least × 40 objective magnification, and preferably × 100 and fluorescence optics with epi-illumination and × 40 and × 100 objectives. Leitz supplies a × 50 water immersion objective, which is particularly useful for observation of routine mycoplasma preparation with Hoechst stain. An automatic camera should also be fitted.

Temperature Recording

Ovens, incubators, and hot rooms should be monitored regularly for uniformity and stability of temperature control. A recording thermometer with ranges from below –50°C to about + 200°C with a resistance thermometer or thermocouple with a long Tefloncoated lead.

Ideally, in addition to the above, recording thermometers should be permanently fixed into your hot room, sterilizing oven and autoclave, and a regular check kept for abnormal behaviour.

Magnetic Stirrer

There are certain specific requirement for *magnetic stirrers*. A rapid stirring action for dissolving chemicals is available with any stirrer but to be used for disaggregation or suspension culture: (1) the stirrer motor should not heat the culture (use the rotating field type of drive or belt drive from an external motor); 2) the speed must be controlled down to 50 rpm; (3) the torque at low rpm should still be capable of stirring up to 10 1 of fluid; (4) it should have more than one place if several cultures are to be maintained simultaneously; (5) each stirrer position should be individually controlled; and (6) there should be a readout of rpm at each position.

Roller Racks

Roller racks are used to scale up monolayer culture. The choice of apparatus is determined by the scale, i.e., the size and the number of bottles to be rolled. This may be calculated from the number of cells required, the maximum attainable cell density, and the surface area of the bottles. A large number of small bottles gives the highest surface area but tends to be more labour-intensive in handling, so a usual compromise is around 125 mm (5 in) diameter and various lengths from 150-500 mm (6-20 in) long. The length of the bottle will determine the maximum yield but is limited by the size of the rack; the height of the rack will determine the number of tiers, i.e. rows of bottles. Although it is cheaper to buy a larger rack than several small ones, the latter alternative: (1) allows you to build up gradually (having confirmed that the system works); (2) can be easier to locate in a hot room; and (3) will still provide accommodation if one rack requires maintenance. The new Brunswick has been found to be reliable and a good size as a starter module if larger scale production is required. Otherwise the Bellco or Luckhams bench-top models may be satisfactory for smaller scale activities.

Pipette Acids and Automatic Pipetting

If a large number of culture is to be made, an automatic pipette such as the Compu-pet or Watson-Marlow will be found to be advantage. (Pipette aids and automatic pipetting are reviewed below). For smaller numbers, Gilson-type pipettors are good for small volumes; Bellco supplies an automated pipette aid which takes regular pipettes. Only the disposable pipette tips of micropipettes need to sterile; the Bellco pipette aid uses regular sterile plugged pipettes.

Mechanical Aids and Automation

Repetitive pipettes have been designed in a variety of patterns and those most suited to tissue culture are the syringe types. They operate either by alternately drawing up and expressing liquid through a two-way valve or by incremental movement of the syringe piston.

Repetitive dispensers can also be mounted on reagent bottles, in which case the culture flask is taken to the pipette rather than vice versa.

All of the repeating pipettes have problems in use resulting from the necessity to autoclave glass syringes, two-way valve, etc. The valves tend to stick (though making these of Teflon helps), syringe pistons deform, or if, Teflon, nay contract due to compression during autoclaving. It is preferable to have a nonsterile metering and repeating mechanism so that only the dispensing element need to sterile. The Bellco automated pipette handle, though it has no facility repetitive pipetting, conforms to conforms to this requirement, as does the Tridak Stopper syringe dispenser.

Automated pipetting can also provided by a peristaltic pump controlled in small increments (e.g., Compu-pet). In these only the delivery tube is autoclaved, and accuracy and reproducibility can be maintained to high levels over ranges from 10 μl up to 10 ml. In addition, a number of delivery tubes may be sterilized and held in stock, allowing a quick changeover in the event of accidental contamination or change in cell type or reagent. Larger pumps (10-15 ml) are also available (e.g. Watson-Marlow). In this volume range, it is also possible to use a simple transfusion device with a graduated reservoir. Graduated reservoirs are less convenient where smaller volumes or greater accuracy is required, although a burette, preferably with a two-way valve, can be used.

The introduction of microtitration trays, has brought with it many automated dispensers, diluters, and other accessories. Transfer devices using perforated trays or multipoint pipettes make it easier to seed from one plate to another, and there are also plate mixers and centrifuge carriers available. The range of equipment is so extensive that it cannot be covered here and the appropriate trade catalogues should be consulted. Two items worthy of note, however, are the Rainin programmable single or multi tip micro pipette and the Coastar Transtar media transfer and replica plating device.

Useful Additional Equipment

Low-Temperature Freezer

Most tissue culture reagents can be stored at 4°C or –20°C, but occasionally some drugs, reagents, or derivatives from cultures may require a temperature of –70°C where most, if not all the water is frozen and most chemical and radiolytic reactions are severely limited. Such a freezer is also a useful accessory for cell freezing. The chest type is more efficient at maintaining a low temperature for minimum power consumption, but vertical cabinet are much less extravagant in floor space. If you do choose a cabinet type, make sure that it has individual compartments (six to eight in a 400 1 (15 ft^3) with separate close-fitting doors, and expect to pay 20% more than for a chest type.

Glassware Washing Machine

A reliable person doing your washing-up is probably the best way of producing clean glassware; but when the amount gets to be too great, or reliable help is not readily available, it may be worth considering an automatic washing machine. There are several of these currently available which are quite satisfactory. You should look for the following principles of operation:

1. Choice of racks with individual spigots over which you can place bottles, flasks, etc. Open vessels such as petri dishes and beakers will wash satisfactorily in a whirling arm spray, but narrow-neck vessels need individual jets. Each jet should have a cushion at its base to protect the neck of the bottle from chipping.
2. The water pump which pumps the water through the jets should have a high delivering pressure, requiring around 2-5 hp, depending on the size of machine.
3. Washing water should be heated to 90°C.
4. There should be a facility for a deionized water rinse at the end of the cycle. This should be heated to 50-60°C; otherwise the glassware may crack after the hot wash and rinse, and should be delivered as a continuous flush and not recycled. If recycling is unavoidable, a minimum of three separate deionized rinses will be required.
5. Preferably, rinse water from the end of the previous wash cycle should be discarded and not retained for the pre-rinse of your next wash. This reduces the risk of cross contamination when the machine is used for chemical and radioisotope wash-up.
6. The machine should be lined with stainless steel and plumbed in stainless steel or nylon pipework.

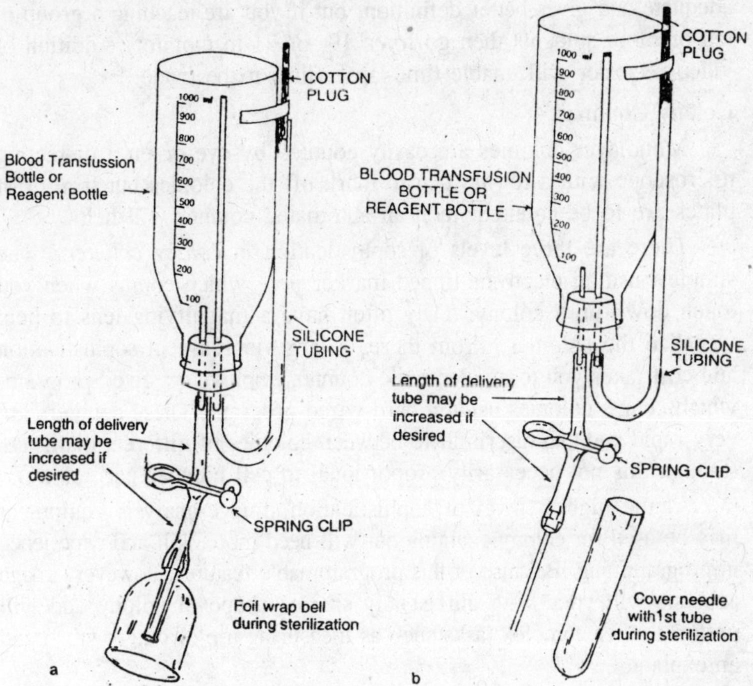

Fig. 2.7. Simple dispensing devices for use with a graduated bottle. (a) With bell used in conjunction with open bottle. (b) With needle for slower delivery via a skirted cap or membrane type closure.

7. If possible, a glassware drier should be chosen that will accept the same racks, so that they may be transferred directly via a suitably designed trolley without unloading.

Betterbuilt make such machines of different sizes with compatible drying ovens.

Closed Circuit TV

Since the advent of cheap microcircuits, television cameras and monitors have become a valuable aid to the discussion of cultures and the training of new staff or students. Choose a high resolution, but not high-sensitivity, camera, as the standard camera sensitivity is usually sufficient, and high sensitivity may lead to problems of over-illumination. Black and white usually gives better resolution and is quite adequate for phase-contrast observation of living cultures. Colour is preferable for fixed and stained specimens. If you will be discussing cultures with a technician or one or two associates, a 12- or 15-in monitor is adequate and gives better definition, but if you are teaching a group of ten or more students then go for a 19- or 21-in monitor. Addition of video recorder will enable time-lapse films to be made.

Colony Counters

Monolayer colonies are easily counted by eye or on a dissecting microscope with a felt-tip pen to mark off the colonies but if a lot of plates are to be counted, then an automated counter will help.

There are three levels of sophistication in *colony counters*. The simplest use an electrode-tipped marker pen, which counts when you touch down on a colony. They often have a magnifying lens to help visualize the colonies. From there, a large increase in sophistication and cost takes you to an electronic counter employing a fixed program, which counts colonies using a hard-wired program. These counters are very rapid and can discriminate between colonies of different diameters (though this not necessarily proportional to cell number per colony).

At the highest level of sophistication, image-analysis equipment may be used for colony counting but will need more skill and experience in programming. Because of this programmable features, however, image analysis will cope with almost any size or shape of colony and will perform other complex tasks such as measuring are of outgrowth round an explant.

Cell Sizing

A dual-threshold cell counter (e.g., the Coulter multisizer) with the facility for pulse-height analysis scans a cell population at a range of threshold settings simultaneously and prints out cell size distributions automatically.

Time-Lapse Cinemicrography

The apparatus may be added to most good-quality inverted microscopes.

Controlled-Rate Cooler

While cells may be frozen by simply placing them in an insulated box at –70°C, some cells many require different cooling rates or differently shaped cooling curves. A programmable freezer enables the cooling rate to be varied by blowing liquid nitrogen into the freezing chamber, under the control of a preset program.

Centrifugal Elutriator

This is a specially adapted centrifuge suitable for separating cells of different sizes. They are costly but very effective.

Flow Cytophotometer

This instrument, also known as *fluorescence-activated cell sorter* (*impulse cytophotometer*, or *cytofluorimeter*) can analyze cell populations and separate them according to a variety of criteria. It has almost unlimited potential but is too expensive to come within the most tissue culture laboratory equipment budgets.

It is always very tempting to purchase new pieces of equipment as they appear on the market, but weigh the advantages that they may offer against the space that they will occupy and what they will cost. Try to be sure also that (1) they will be of lasting benefit and (2) you and others will want to use them.

Consumable Items

This includes general items such as pipettes, culture flasks, ampules for freeing, centrifuge tubes (10-15 ml, 50 ml, 250 ml; Sterilin, Corning), universal containers disposable syringes and needles (21-23 g for withdrawing fluid from vials, 18 g for dispensing cells), filters of various sizes for sterilization of fluids, surgical gloves, and paper towels.

Pipettes

These should be "blow out" and wide tipped for last delivery, graduated to the tip with the maximum point of the scale at the top rather than at the tip. Disposable pipettes can be used but are expensive and may need to be reserved for holidays or crises in wash-up or sterilization. Pipettes to be re-used are collected in pipette cylinders or hods, one per work station.

Pasteur pipettes are best regarded as disposable and should not be discarded into pipette cylinders but into secure glassware waste.

Culture Vessels

Choice of culture vessels is determined by (1) the yield (cell number) required; (2) whether the cell is grown in monolayer or suspension; and (3) the sampling regime, i.e., are the samples to be collected simultaneously or at intervals over a period of time.

"Shopping around" will often in a cheaper price, but do not be tempted to change too often and always test a new supplier's product before committing yourself.

Care should be taken to label "sterile," nonsterile", "tissue culture" grade, and "non-tissue culture" grade plastics clearly, and preferably they should be stored separately. Glass bottles with flat sides can be used instead of plastic provided a suitable wash-up and sterilization service is available.

Cleaning Procedures

Glassware

All glassware should be placed in water immediately after use. In this way, proteinaceous deposits are prevented from driving out and this greatly facilitate cleaning. The collected glassware is prepared by removing cotton-wool plugs, etc., and by rinsing with water. It is then subjected to whatever cleaning method is employed. A great many procedures have been recommended and these can be divided into three general groups:

1. Clearing with alkalies.
2. Cleaning with detergents.
3. Cleaning with oxidizing acids.

Alkalies

Many different alkalies have been used. The most popular are soft soap, sodium triphosphate, sodium carbonate and sodium metasilicate. There is little to choose between them but sodium metasilicate is probably the most commonly used. It is considered to be particularly suitable since it introduces no foreign ions to the glassware and, if a monolayer of metasilicate remains, it is deposited as glass on neutralization.

The usual practice is to prepare the alkali solution in a concentration about 100 times stronger than ultimately required. The glassware is boiled in water to which this stock solution has been added. The glassware is subsequently treated with dilute acid and then rinsed with water, dried and sterilized.

A detailed description follows of the use of sodium metasilicate as a cleaning solution.

The solution required are as follows:

1. Sodium metasilicate solution ($\times 100$).

Sodium metasilicate	360 g.
Calgon or calgolac	40 g.

Dissolve in a gallon of water and filter. For use add this as

1. Part in 100 to tap water.
2. Normal hydrochloric acid.
3. A supply of deionised water.

It is desirable to have several containers so that glassware can be transferred from one to another. An electrically heated domestic wash boiler is very suitable for boiling up glassware and large fish tanks or polythene baby baths for soaking glassware. The procedure is as follows:

1. The glassware is scrubbed in water to remove debris.
2. It is transferred to a boiler containing metasilicate solution. This is brought to boiling point, maintained there for 20 minutes and allowed to cool.
3. The glassware is removed from the boiler and rinsed thoroughly two or three times with ordinary tap water.
4. It is then placed in a bath containing dilute hydrochloric acid (N/100). After soaking in this for some hours it is given a final rinse with tap water followed by one with deionised water and is allowed to drain. Subsequently, it is dried in an oven and prepared for sterilization.

All glassware is treated in exactly the same way with the exception of the following items:

Pipetties

Pipettes are collected in jars containing water and before cleaning cotton-wool plugs are removed. They are rinsed thoroughly in a Pipettes-washer of the siphon type and transferred to a large jar containing metasilicate solution. After soaking in this for 24 hours they are rinsed quickly and transferred to another jar for N/100 hydrochloric acid. Finally, after another quick rinse in tap water, they are placed in 95 per cent alcohol, drained, dried and sterilized.

Paraffined objects

Depression slides and other objects which have received coating of paraffin must be handled separately since the melted paraffin will form a layer on the surface of the washing solution which will subsequently be deposited on all other glassware. Most of the paraffin is scraped off with a knife or razor blade and the slides are placed in a basin or beaker with a larger pad of gauze in the bottom to prevent damage. They are then covered with water or metasilicate solution which is brought to the boil. The paraffin is floated off by flooding the vessel with water while the surface layer is poured off. The slides, thus partially cleaned, can then be carried through the normal procedure for glassware.

Siliconed objects

Glassware which has been treated with silicone is best handles separately since the silicone may be disseminated to all the other material being cleaned simultaneously. Silicone can be removed by heating in 0.5N NaOH or by treating the glassware with alcoholic KOH. In either case it is best to float off the scum in the same way as was described above for the cleaning of paraffined objects.

Cover slip

Cover slips have to be handled separately because of their small size and fragility, but the same general principles of cleaning are applied. They are dropped one by one into beaker of boiling metasilicate solution then removed, rinsed, treated with 95 per cent alcohol and dried. Finally, they are polished between two boards covered with cloth, transferred to Petri dishes and sterilized by dry heat.

Detergents

Detergents are quite widely used in the cleaning of glassware. They have advantage that no boiling is required. On the other hand, they are difficult to remove and very through rinsing indeed is required. Also, many detergents are toxic to cells so that care has to be taken in choosing tissue one. A detergent which has been evolved specially for cleaning tissue culture glassware is that known as Microsolve and manufactured by Microbiological Associates of Bethesda, U.S.A. This detergent is particularly non-toxic. Other suitable detergents include 7X, Haemosol and Stergene. The cleaning procedure is exactly the same as with alkalies except that boiling is unnecessary and the hydrochloric acid rinse can be omitted.

Oxidising Acids

Many workers insist on the use of strongly oxidizing acids for cleaning glassware, but it is rarely necessary to use them for biological materials. The acids most commonly used are hot *sulphuric acid*, *chromic acid*, and *nitric acid*. They have the disadvantage that a certain amount of danger is involved in their use. The technical staff have to be well-trained and reliable. On the other hand, they certain remove all organic material and probably etch the glass less than alkalies. However, they are difficult to remove and in the case of chromic acid it has been shown that a monolayer tends to remain on the glass even after very prolonged rinsing. Nevertheless, chromic acid is frequently used. It is best prepared as follows: 40 g. of potassium dichromate are placed in a 5-litre Pyrex beaker and a little water added to dissolve it. Concentrated sulphuric acid is then added to make up 1 litre. The solution should be yellow-brown in colour. The development of a green colour indicates that the chromic acid has become reduced and is then useless and should be discarded. Chromic acid is highly corrosive and must be used with great care. It is very often adequate to cover the inside the glassware with a thin layer of the acid and for this purpose the simple apparatus illustrated can be used. On pumping with the hand bellows a jet of chromic acid is forced up inside the funnel into the vessel to be cleaned. The acid drains back into the flask when the hand bellows is released. It is usually adequate to leave glassware in contact with chromic acid at room temperature. Only very occasionally it is necessary or desirable to use chromic acid at a higher temperature and this must be done with the greatest care.

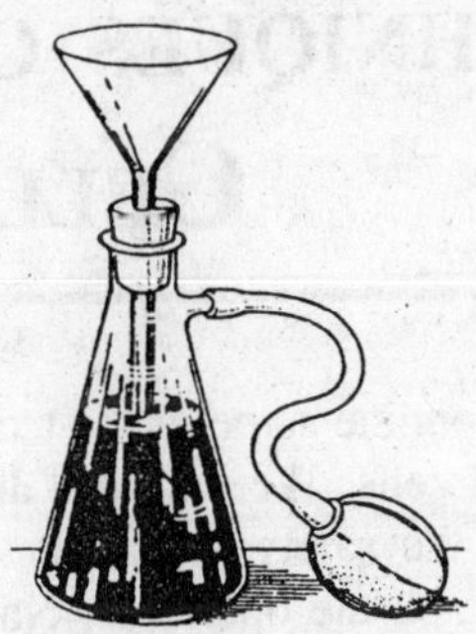

Fig. 2.8. Apparatus for coating the inside of vessels with a layer of chromic acid. On pressing the hand bellows a jet of chromic acid is forced into a vessel held over the filter funnel.

Nitric acid is frequently used for cleaning small objects, particularly cover glasses. The same remarks apply to nitric acid as to chromic acid. The cover glasses are usually dropped into nitric acid and then removed and rinsed very thoroughly indeed. Note that nitrate can occur physiologically and is probably a very much less toxic ion than chromate so that it is much less objectionable.

At the National Cancer Institute in the United States, large steam kettles filled with concentrated sulphuric acid are used for cleaning all the glassware. Very few other laboratories use this technique, however.

One kind of apparatus which is best treated with oxidizing acids is sintered glass filters. They are cleaned with concentrated sulphuric acid to which a few crystals of sodium nitrate and sodium chlorate have been added. The mixture is allowed to percolate through the subsequently the filters are rinsed very thoroughly with a very large volume of distilled or deionised water before drying and sterilizing.

3

Techniques of Animal Cell Culture

In this chapter we will outline some of the techniques commonly used in the culture of animal cells. We will also discuss the importance of aseptic techniques and the safety precautions that must be taken to protect the cell culture and the operators. Note that we use the term aseptic techniques to cover the whole range of procedures used to prevent contamination of cultures by unwanted micro-organisms (contaminants). In later chapters, we will discuss aspects of large scale cell culture, characterization of cell lines and their preservation. The emphasis here is to explain how to establish and maintain cell lines from whole tissue in a simple laboratory setting.

Setting up the Laboratory

Perhaps the biggest pre-occupation of the cell culture biologist is how to prevent contamination of the tissue cultures. The media used to grow the cells also provide excellent nutrition for unwanted organisms. Contamination of cell cultures by bacteria, fungi and mycoplasma often results in the loss of a great deal of time and money. An additional problem is the cross contamination of cells from other cell lines. Before we examine how to ensure contamination is kept at a minimum we need to briefly consider how we would detect contamination.

Regularly checking aliquots from culture flasks using a microscope can reveal some types of contamination, for example, yeasts and fungi can be seen easily. Bacteria Gill be Gram stained or plated out on blood agar plates and general cloudiness of medium also indicates yeast or bacterial contamination. Mycoplasmas are more insidious, the medium is not cloudy, the organisms cannot be seen under an ordinary microscope and, most mycoplasma species are very difficult to grow on agar plates. Testing for mycoplasma is time consuming but must be done regularly because cross contamination is very easy. DNA stains or molecular probes may be used to detect their presence in cells, those cells which turn out to contain DNA in the cytoplasm must be discarded.

Checking Media Ingredients and Glassware

Much of the contamination comes from the cell culture medium and its components or from inadequately cleaned glassware. Routine sterility checks on the medium, serum and nutrients are recommended. To do this, a small aliquot of the medium should be incubated at room temperature and another similar aliquot simultaneously incubated at 37°C for up to one week before the medium is used for culturing cells. If these small samples are found to be contaminated, the medium should be autoclaved and discarded.

The use of antibiotics in the medium helps to contain the problem to some extent but it should never be seen as an alternative to good aseptic technique and careful monitoring.

Nature of The Work Area

Where possible, a separate room should be made available for clean cell culture work. This room should be free of through traffic and, if possible, equipped with an air flow cabinet which supplies filtered air around the work surface. A HEP A (High Efficiency Particle Air Filter) filtered air supply is desirable but not always affordable. Primary animal tissue and micro-organisms must not be cultured in or near the cell culture laboratory and the laboratory must be specifically designated for clean cell culture work. Clean laboratory coats should be kept at the entrance and should not be worn outside of this laboratory and brought back in.

All work surface, benches and shelves and the base of the airflow cabinets must be kept clean by frequent swabbing with 70% alcohol or an alternative disinfectant. If an airflow cabinet cannot be provided, the culture work may be done on a clean bench using a bunsen burner to create a sterile 'umbrella' under which the work can be done.

Aseptic Techniques

You have probably already some experience of aseptic techniques from culturing micro-organisms. Similar techniques may be used to transfer cultures of animal cells. The basic rules for aseptic techniques which should be used even if an airflow cabinet is available include:

1. If working on the bench, use a bunsen flame to heat the air surrounding the bunsen. This causes the movement of air and contaminants upwards and reduces the chance of contamination entering open vessels. Open all bottles and perform all manoeuvres in this area only;
2. Swab all bottle tops and necks with 70% alcohol to clean them before opening;
3. Flame all bottle necks and pipettes by passing very quickly through the hottest part of the flame. This is not necessary with sterile, individually wrapped, plastic flasks and pipettes;
4. Avoid placing caps and pipettes down on the bench; practice holding bottle tops with the little finger while holding the bottles for pouring or pipetting;
5. Work either left to right or vice versa, so that all material to be used is on one side and, once finished, is placed on the other side of the bunsen burner. (This may also stop the operator using the same reagent twice!);
6. Manipulate bottles and flasks carefully. The tops of bottles and flasks must not be touched by the operator. Touching of open vessels should also be prevented when pouring. If necessary practice pouring from one container to another keeping a distance of 5 mm between the two vessels;
7. Clear up spills immediately and always leave the work area clean and tidy. Dispose of glassware in appropriate bins and discard used plasticware in marked polythene bags for autoclaving or incineration. All glassware or plastic ware used for infectious work must always be autoclaved before incineration. Re-usable glassware should be immersed in disinfectant whilst awaiting transfer to an autoclave.

The rules we have listed above are by no means exhaustive. The ways in which they are implemented are, however, slightly different in different laboratories. Learning to apply these rules depends upon gaining 'hands on' experience within a laboratory. You should not attempt to carry out culture transfers without being shown how to do it properly by an experienced operator.

Basic Equipment Used in Cell Culture

In addition to an *airflow cabinet* and *benching* which can be easily cleaned, the culture laboratory will need to be furnished with an incubator or hot room to maintain the cells at 30-40° C. The incubation temperature will depend on the type of cells being cultivated. Insect cells will grow best at around 30°C while mammalian cells require a temperature of 37°C. It may be necessary to use an incubator which has been designed to allow CO_2 to be supplied from a mains supply or gas cylinder so that an atmosphere of between 2-5% CO_2 is maintained in the incubator.

A refrigerator or cold room is required to store medium and buffers. A freezer will be needed for keeping pre-aliquoted stocks of serum, nutrients and antibiotics. Reagents may be stored at a temperature of –20°C but if cells are to be preserved it may be necessary to provide liquid nitrogen or a –70°C freezer.

A microscope with normal Kohler illumination will be needed for cell counting. An inverted microscope will also be needed for examining flasks and multiwell dishes from underneath. Both microscopes should be equipped with a x10 and a x20 objective and it may be useful to provide a x40 and a x100 objective for the normal microscope Additional features such as a camera, adaptor and attachments and UV facility may also be required for some purposes. A waterbath and a centrifuge with sealed buckets is also necessary. Additional requirements include counting chambers (Improved Neubauer or a Coulter counter for counting cells, glass or plastic cell culture flasks, graduated pipettes of various sizes, centrifuge tubes and universal containers, disposable pasteur pipettes, rubber bulbs or automated pipetters for use with pipettes and precise calibrated pipetters, e.g. Gilsons, for measuring small volumes from 1-1000 microlitres. More specialized equipment may be needed for particular experiments.

Culturing Cells

Sources of Tissue for Culture

A wide variety of cells from insects, fish, mammals and humans may be grown *in vitro* some as primary cultures only but many as secondary and continuous cell lines.

Early attempts at culturing tissues relied upon the explantation of whole tissue or organ which could be maintained *in vitro* for only very short periods. Nowadays it is more usual to grow specific cell types from tissues, although there are still some situations where it is necessary to grow a whole organ (or a part of it).

Some cell functions such as respiration, proliferation and gene transcription can proceed normally even when the cells are isolated from the parent organ. Other functions, such as the production of hormones or a response to external stimuli, are dependent on the interaction of several cell types within the organ.

In organ culture, whole organs or parts of organs in culture are maintained in order to maintain the *in vivo* interactions of the different cell types within the tissue *in vitro*, so that the effect of exogenous stimuli on the whole organ can be studied. For example, pieces of guinea pig or rabbit gut are used to test the effect of atropine-like drugs on the smooth muscle of the gut because nerve synapses and muscle tissue have to work together to produce a measurable response to stimuli.

There are disadvantages to organ cultures. Organs cannot be propagated so each piece of tissue can only be used once, which makes it difficult to assess the reproducibility of a response. And, of course, the particular cells of interest may be very small in number in a given piece of tissue so the response produced may be difficult to detect and quantify. It may not be possible to supply adequate oxygen and nutrients throughout the tissue because of the absence of a functioning vascular system, so necrosis of some cells occurs fairly rapidly. This problem may be ameliorated to some extent by keeping the organ in stirred cultures or in roller bottles which alternately provide air and soluble nutrients.

Sources of Cells for Culture

Liver, lung, breast, kidney, skin and bladder tissue from animals and humans have been used as sources of epithelial cells for tissue culture. Muscles, bone and cartilage tissue and neural cells can also be grown *in vitro* as can blood cells and many types of tumours cells.

Embryo-derived cell lines are easier to establish and sub-culture than tissue derived from newborn or adult animals. With adult tissue it is more difficult to obtain viable proliferating cells because the onset of differentiation is already under way. There is also an increase in fibrous connective tissue which makes disaggregation more difficult and there is a significant reduction in the undifferentiated proliferating cell pool.

Many of the cell lines in common use are widely available from cell banks and type culture collections. Some very well known cell lines include the HeLa Line (from a cervical carcinoma) and MRCS (a fibroblast line developed by MRC laboratories).

Occasionally, however, it is necessary to use primary or low passage cells and these can be prepared by obtaining tissues from laboratory animals or from tissue banks. The whole tissue needs to be disaggregated to produce cells which can be cultured.

Preparation of Primary Cultures

It is not always necessary to disaggregate tissue before culturing. Some embryonic tissue can be cultured simply by leaving the whole tissue on the flask surface and individual cells will simply grow out from the whole tissue and proliferate. After a few days the original tissue can be removed and the culture medium replenished to allow the new cells to continue growing. This method works for some tissues but it is not suitable for growing specific cells from a piece of tissue.

Tissues that do not require enzymatic disaggregation

Tissuc for primary cxplant can be treated as follows:

1. Excess blood is removed by rinsing with a sterile *balanced salt solution* (BSS). If the tissue is to be transported it can be kept in BSS or medium;
2. Unwanted material such as fat and cartilage is cut off and the rest is chopped finely using a pair of crossed scalpels. It is important to make clean cuts, the use of tearing actions or scissors may damage the cells;
3. The cell suspension is transferred to a sterile 50 ml centrifuge tube together with the buffered saline and the cells are allowed to settle out;
4. The medium is carefully pipetted off and the pellet is washed in fresh BSS. The cells are allowed to settle again or the suspension is gently centrifuged at about 1000 rpm for 5 minutes;
5. The cell pellet is resuspended in 10-15 ml of medium and the suspension is aliquoted into 2 or 325 cm^2 flasks. If the pieces of tissue are still quite large the suspension ma) be passed through a sieve before culturing, the cultures are incubated at the appropriate temperature for 18-24 hours. (Note that in describing the cultivation of adherent cells, it is usual to describe the size of the vessels in terms of surface area rather than volume);
6. If the pieces of tissue have adhered, the medium should be changed weekly until a significant outgrowth of cells has occurred;
7. The original tissue pieces which show outgrowth can then be picked off and transferred to a fresh flask;
8. The medium in the original flask is replaced and the cells cultivated until they cover at least 50% of the surface available for growth. They can then be sub-cultured if required.

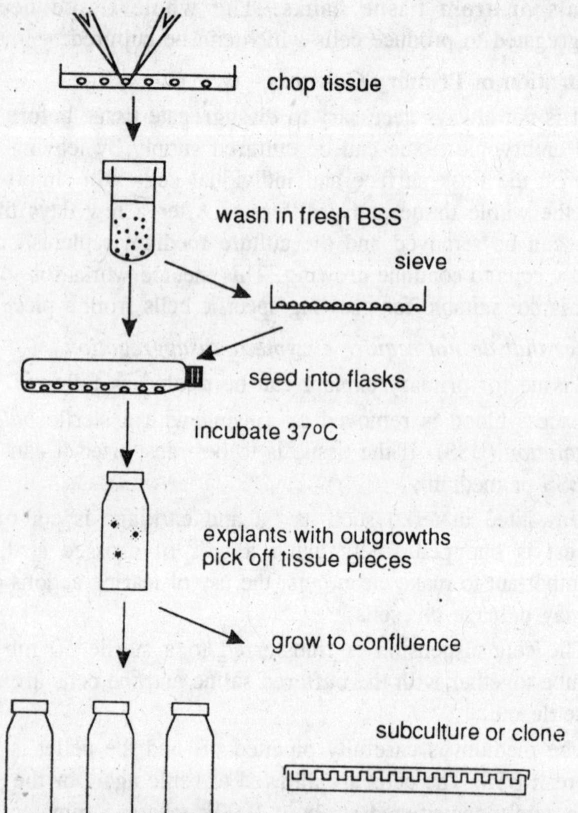

Fig. 3.1. Disaggregation of tissue and primary culture.

All steps must, of course, be done in such a way as to reduce the chances of the tissue, media and vessels becoming contaminated. Thus all dissection tools, centrifuge tubes and reagents need to be sterilized before use.

This method is useful for small pieces of tissue such as skin biopsies. Fibroblasts, glial cells, epithelium and myoblasts all migrate out of the tissue very successfully. Some selection occurs because not all cells will adhere to the same extent and some cells rapidly outgrow the others.

Tissues that require enzymatic disaggregation

The tissue may be kept whole or may be cut into small pieces as described above. However, densely packed tissue is more difficult to digest and prolonged contact with digestive enzymes such as trypsin often causes destruction of viable cells. Large pieces of tissue for enzymatic disaggregation should be cut into smaller pieces as for the primary explant and digested with trypsin for about 30 minutes at 37°C. Then the trypsin is neutralized by the addition of serum. The amount of trypsin that is required to achieve satisfactory release of cells is tissue-dependent and also depends upon the activity of the enzyme preparation. Trypsin suppliers often provide useful guidance for the use of their own preparations. Typically 0.25% w/v trypsin solution is satisfactory.

Trypsin is a proteolytic enzyme which hydrolyses proteins. By adding a lot of protein (in the form of serum), the enzyme begins to hydrolyse these proteins rather than the proteins which bind cells together. What we are really doing is 'diverting' the enzyme to an alternative substrate so that it is no longer available to attack the extracellular matrix.

Thus once *trypsinization* is complete, medium containing serum should be added to neutralize enzyme activity. If trypsin is used to disaggregate whole tissue, the dissociated cells should be harvested after about 30 minutes and washed to remove trypsin. The remaining whole tissue may be retrypsinized. An alternative method is to soak the piece of tissue in cold trypsin at 4°C overnight. The cold trypsin penetrates throughout the whole tissue but will have minimal activity at this temperature. The tissue can be disaggregated by increasing the temperature to 37°C for 30-40 minutes, then enzyme activity is neutralized by the additions of medium with serum as before.

Collagenase and versene (phosphate buffered saline + EDT A) are also used to disaggregate tissues. Both are gentler in action than trypsin but expensive because larger amounts are required. Collagenase or versene can also be used together with trypsin to allow trypsin to be used at lower concentrations.

A variety of other hydrolytic enzymes (for example pronase) have also been employed to disaggregate tissues in some laboratories.

Once disaggregation is complete, the cells can be washed in fresh BSS or medium and seeded out in 24 cm^2 flasks. The cell layer can be supplied with fresh medium after 48 hours. This culture is referred to as a primary cell culture.

Removing Non-viable Cells from the Primary Culture

If the primary culture is of anchorage-dependent or adherent cells, the non-viable cells can be removed by pouring off the medium and rinsing the cell layer with buffer before adding fresh medium. Cell which are not viable will not be able to adhere to the substratum. If the cells are to be cultured in suspension, the non-viable cells gradually become diluted as the viable cells proliferate. If it is necessary to remove the non-viable cell from the suspension, the cell may be layered onto Ficoll or 'lymphoprep' and be centrifuged at 2000 rpm for 15-20 minutes. The non-viable cells will sink to the bottom and the viable cells can be collected from the medium-Ficoll interface.

MAINTAINING THE CULTURE

If a primary culture is not to be used as such, it may be sub-cultured to produce a *cell line*. Cell lines may be of a very limited lifespan or they may be passaged several times before the cells become senescent. Some cells, such as macrophages and neurones do not divide *in vitro* and can only be used as primary cultures.

Sub-culturing from Primary to Secondary Cell Culture

A primary culture contains a very heterogeneous population of cells from the original explant. Some of these cells will die, some will fail to grow, others will grow quickly and become the dominant cell type present. On *sub-culturing* the primary cell culture, the dominant types will become even more dominant. We can, therefore, foresee that sub-culturing enables us to produce more homogenous cell populations.

Producing a cell line has certain obvious advantages. A homogeneous population of characterized cells can be grown to a large scale, replicates are uniform and this makes designing experiments much easier. There are, however, disadvantages. In establishing a cell line only those cells best suited to the *in vitro* conditions are selected for. These cells may lose some of the differentiated characteristics they had while growing *in vivo* and are prone to genetic instability, particularly if they divide rapidly.

We really have only two main options here. First, of course, we must choose the appropriate tissue in the primary explant stage. It is no good attempting to produce a particular epithelial cell line if we use tissues which do not contain the appropriate cell type. Secondly, not all cells will grow equally well in the same medium. So, in principle, by selecting our medium carefully we may provide the condition most suited to our cells of interest.

Propagating a Cell Line

Once a cell line is established, it needs to be propagated in order to produce sufficient cells for characterization and storage, as well as for particular experiments.

A cell line is given a name or code which identifies its source (for example HuT, Human T cells) and, if more than one line was developed from the same source, a cell line number is also given (for example HuT 78). If cells from this line are cloned, then a clone number need to be given, e.g. HuT 78 clone 6D5.

If the cell line is likely to be viable for only a few sub-culturings or generations each generation should be noted, for example HuT 78 clone 6D5/2 or HuT 78 clone 6D5/3.

Once a culture is confluent (i.e. the cells cover 60-70% of the growth surface available) can be transferred into a *holding* or *maintenance medium* which provides just enough nutrients to keep the cells alive and healthy but reduces the replication rate so that the cells do not overgrow. The usual method is to use the same basic medium and reduce the amount of serum it contains.

If cells are required for experiments or storage the contents of the flask can be split or divided to seed 2-4 new flasks depending on the vigour of the cell growth.

The content of a flask of cells grown in suspension is very simple to split. This is done in the following way:

1. Stand the flask upright for 1-5 minutes to allow cells to settle;
2. Aseptically remove as much medium as possible without disturbing the cells. A 10 ml graduated pipette is best. Discard used medium into a waste pot for autoclaving;
3. Resuspend cells in the remaining volume, measure and divide between the required number of flasks;
4. Top up each flask with fresh growth medium and incubate as normal;
5. Record the cell line code, clone number and the passage number.

Cells which grow as monolayers adhering to the flask surface are a little more difficult to split.

The steps are:

1. Remove medium by pouring it off aseptically;
2. Rinse the monolayer with prewarmed (37°C) phosphate buffered saline (PBS) BSS; pour the medium off into discard pot;
3. Add 10 ml of a 0.25%.(w/v) solution of trypsin in saline. Tilt the flask so that all the cells are covered with trypsin. Other enzymes may also be used in the same way;
4. Pour off the trypsin and incubate the flask at 37°C for 1-10 minutes. Cells which very susceptible to trypsin should be checked after 1 minute;

5. Once the cells have begun to round up and are sliding off the flask surface (it may help to tip the side of the flask gently), the trypsin must be neutralized by addition of 5-10 ml of medium containing serum;
6. Wash all the cells down from the sides of the flask and aspirate gently using a pipette. This should break up any aggregates and allow for easier cell counting if required. A void creating froth as this can lead to contamination and cell damage;
7. A small aliquot (0.2 ml) of the cell suspension may be removed for counting if required. Otherwise, measure the total volume of the cell suspension and divide between the required number of flasks and top up and incubate as before;
8. Remember to record the cell line code, clone number and the number of times it has been divided (the passage number).

So if a flask of fibroblast cells is divided into two new flasks as well as the original flask the following type of labelling system can be used.

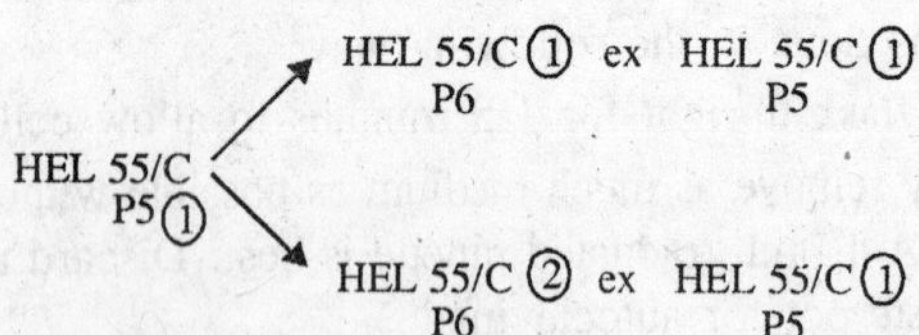

Where flasks 1 and 2 are derived from the single flask P5 which represents the passage number. *Note* "ex" refers to the flask of origin.

The intervals between sub-culturing or changing medium will depend on the cell line and the rate at which it grows. Rapidly dividing cell lines such as HeLa and VERO need to be sub-cultured at least once a week with a medium change in between. Fibroblast HEL cells may be sub-cultured once every 10-14 days. The rate of growth of fibroblasts may be increased, if more frequent sub-culturing is required, by increasing the serum concentration in the medium.

Quantitation of Cells in Cell Culture

For properly run experiments, it may be necessary to count the cell numbers before, after and even during the experiment. Day to day maintenance of cell lines also requires quantitative assessment of cell growth so that optimum cell densities for sub-culturing and storing can be determined.

We can divide the methods available for determining cell growth into two sub-groups. These are:

1. Direct methods;
2. Indirect methods.

In the *direct method*, cell numbers are determined directly either by counting using a counting chamber or by using an electronic particle counter. In the *indirect methods* measurement of some parameter such as DNA content or protein content related to cell number is used as the method of estimating biomass.

Direct Methods for Quantitation of Cells in Culture

Counting chambers

The most commonly used device is the Improved Neubauer haemocytometer originally designed for counting blood cells. It consists of a thickened slide with a central chamber of known depth. A grid is etched out of the silvered chamber bottom.

The counting chamber is prepared and loaded with a suspension of single cells for counting. It is important to aspirate the cell suspension adequately before loading the chamber in order to break up clumps of cells which are difficult to count accurately. The counting chamber is examined under a microscope using a x10 objective. The cells in the grid are counted.

Since the cells are distributed randomly over the grid then, in principle the more cells we count the more accurate our result will be. A good 'rule of thumb' is to use the following relationship. If we count N cells then our error is $\sqrt{N}$. Thus if we count 49 cells then our error will be of the order of 14%, whilst if we count 4900 cells our error is 1.4%. In practice, because of errors elsewhere in the process (for example uncertainties in eluting all cells by trypsinization), it is pointless attempting to count a very large number of cells. Usually if we count 100 or more cells this will be sufficient for most purposes.

Let us turn this into a practical scheme. First we load the chamber with cells. We can then count all of the cells within the 1 mm square grid. The cell concentration ml-1 of suspension can be calculated using the relationship

$$C = \frac{n}{V}$$

where C = cell concentration
n = number of cells counted
V = volume (ml) represented by the grid.

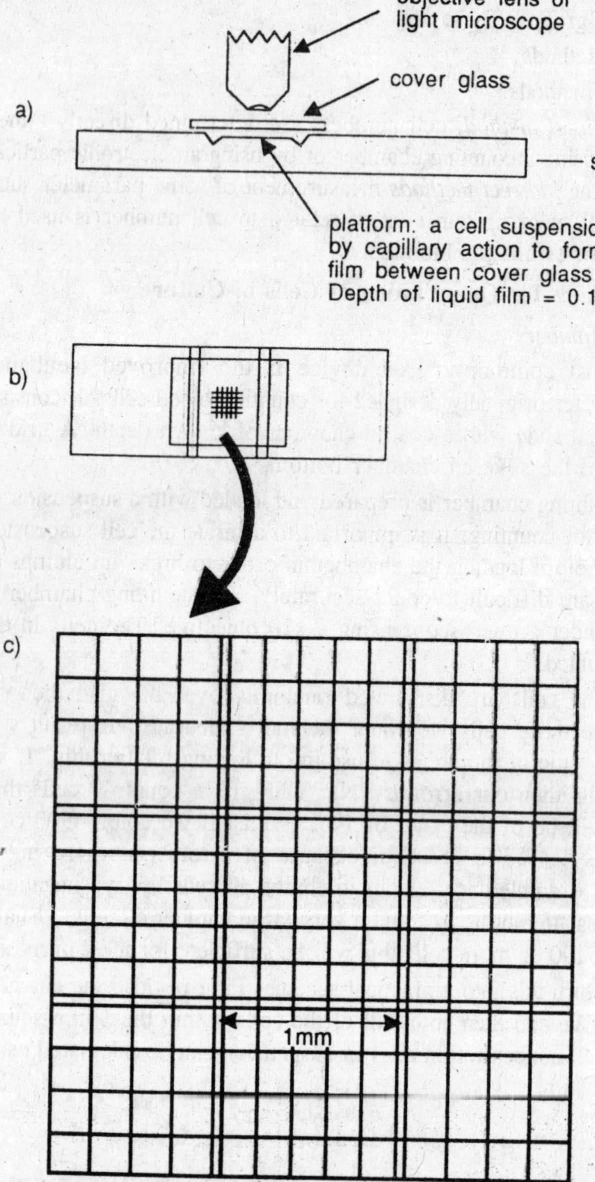

Fig. 3.2. A typical counting chamber used to determine cell numbers. Note that different manufacturers use slightly different formats. Some have two sets of grids per slide and the individual blocks of separates are separated by three, rather than two, score lines.

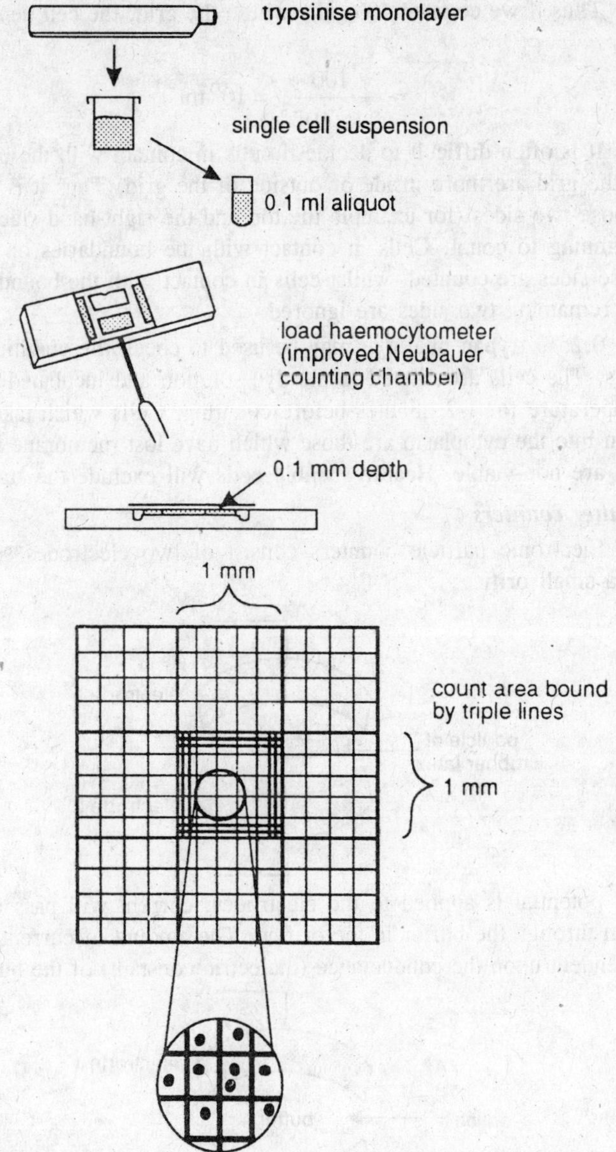

Fig. 3.3. Preparing samples and loading haemocytometers for cell counting.

The volume of suspension over the grid is:

$1 \times 1 \times 0.1\ mm^3 = 0.1\ mm^3 = 0.1 \times 10^{-3}$ ml.

Thus if we counted 100 cells within the grid, the cell density will be:

$$= \frac{100}{0.1 \times 10^{-3}} = 10^{6}\,\text{ml}^{-1}$$

It is often difficult to decide if cells in contact with the boundary of the grid are more inside or outside of the grid. Thus it is usual to choose two sides (for example the top and the right-hand side) before beginning to count. Cells in contact with the boundaries on each of these sides are counted, whilst cells in contact with the boundaries on the remaining two sides are ignored.

0.25% trypan blue dye may be used to check the viability of the cells. The cells are diluted in the dye solution and incubated at room temperature for 1-2 minutes before counting. Cells which take up the stain into the cytoplasm are those which have lost membrane integrity and are non-viable. Healthy, viable cells will exclude the stain.

Coulter counters

Electronic particle counters consist of two electrodes separated by a small orifice:

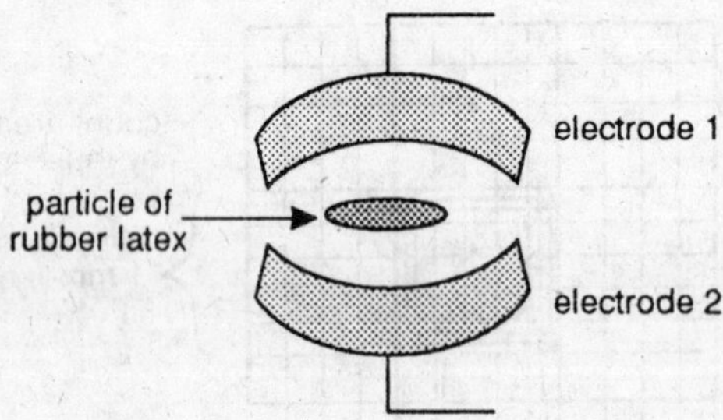

If a potential is applied to the electrodes, current will pass between them through the buffer in the orifice. The amount of current will be dependent upon the conductance (dielectric constant) of the buffer.

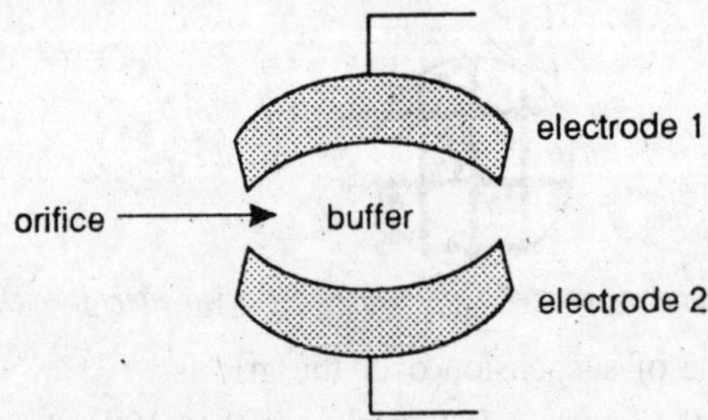

As the particle enters the orifice, the conductance of the solution between the electrodes would be reduced. Thus the current flowing would be reduced and this could be detected electronically.

The size of the change in current flow depends on the size of the particle and the difference in the dielectric constant (conductivity) of the particle and the suspending buffer. This is the principle upon which cell counting using a Coulter counter is carried out.

Cells in suspension are drawn through a fine orifice in the Coulter counter and, as each cell passes through, it produces a change in the current flowing across the orifice. Each change is recorded as a pulse and it is these pulses that are sorted and counted. The size of the pulse is proportional to the volume of the particle passing through, so signals of varying size are produced. The pulse height threshold can be set, therefore, to eliminate electronic noise and weak pulses produced by debris.

Electronic counters, of which the Coulter counter is the most widely used, provide rapid results, but high cell numbers are required to give an accurate count. Another disadvantage is that they cannot distinguish between dead and viable cells and clumps of cells may register as a single pulse thus leading to inaccurate counts.

Indirect Methods for Determining Cells in Culture

Other methods of quantitation such as radioisotope labelling and estimation of total DNA or protein are used less frequently. They are useful when cells are grown in microwell plates or as hanging drop cultures. The cells can be mixed *in situ*, stained and counted by eye under a microscope. DNA and protein assays are inaccurate, particularly if cells are multinucleated. They do not distinguish viable and non-viable cells.

Cloning and Selecting Cell Lines

We have mentioned before that most primary explant cultures consist of very heterogeneous populations of cells, and often when sub-cultured the cells which do not grow under the given conditions are lost. Slow growing cells may rapidly become overgrown by faster growing cells.

Specific cell types can be separated by physical separation techniques or by cloning in order to produce pure cell lines of single cell types. In 1966 Coon and Cahn separated cartilage and pigment-producing cell strains by cloning. The cloned cell lines retained their specialized functions over several generations. In 1978 Clark and

Pateman cloned a primary culture of Chinese Hamster Liver to produce a cell line of Kupffer cells.

There are a number of ways to produce a cloned cell line. These include:

1. Cloning by dilution;
2. Cloning by interactions with substrate;
3. Cloning by selective detachment.

Cloning by Dilution

Cells from a heterogeneous culture are trypsinized or agitated to produce a suspension of single cells. The suspension is diluted to give between 10 and 100 cells ml^{-1} by serial dilution; for example a 1 ml aliquot of a suspension containing 1000 cells ml^{-1} is diluted with 9 ml of medium to give 100 cells ml^{-1} and so on. The diluted cell suspension is seeded into microwell plates with the appropriate growth medium and supplements and allowed to grow for 2-3 weeks. The cells from each well are transferred to flask and grown for characterization. To some extent specific cell types may also be selected by adjusting the components of the medium to encourage or discourage particular cell types.

Diluted cell suspensions can also be seeded into sterile petri dishes or flasks and grown until small colonies begin to form. The individual colonies can be physically isolated from each other with cloning rings made of porcelain, stainless steel and nylon. The rings, placed around single colonies, simply prevent different colonies growing into one another.

It is often necessary to adjust the normal growth medium components when growing clones. Most cells require a minimum number of other cells in their vicinity in order to grow. Cells grown at very low densities may require nutrients from the growth medium which would normally be produced as soluble factors by surrounding cells (for example cytokines such as interleukin 2 which is necessary for T cells to proliferate). These cell-derived products may be present in negligible concentrations in low density cultures.

Problems of low density cultures may be overcome by growing the cells in very small volumes e.g. capillaries or hanging drop cultures. Feeder layers of, for example irradiated mouse cells, can also be used, though the disadvantage with this method is that the cloned cells then need to be separated from the feeder cells. However, irradiated feeder cells do not divide in culture.

Carefully selected foetal bovine serum, enriched medium or conditioned medium (medium taken from cultures of similar cells which have been grown to 50% confluence, filtered and used to mix with the growth medium for cells to be cloned) may be used to provide essential cell-derived growth factors. The addition of hormones such as hydrocortisone and insulin may also help to improve growth efficiency.

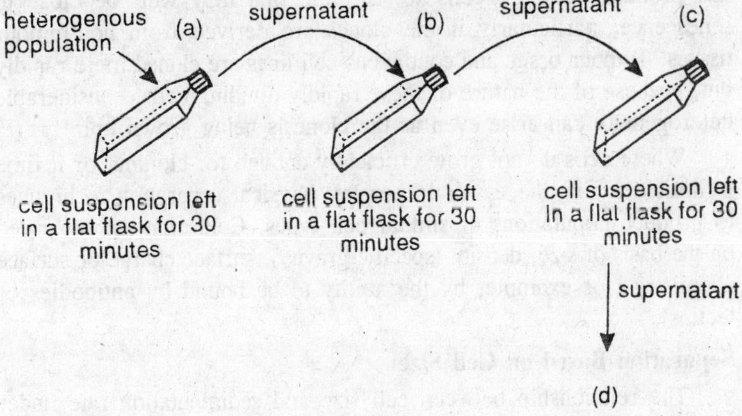

Fig. 3.4. Figure showing a scheme for separating specific cell types from a heterogeneous population.

Cloning by Interaction with a Substratum

Different cell types attach to substrata at different rates. This feature can be used to separate different types of cells from a heterogeneous mixture. For instance, if a primary cell suspension is seeded into one flask or petri dish and incubated for 30 minutes and then transferred to a second flask for another 30 minutes and so on, the most adhesive cells will be found in the first flask and the least adhesive ones in the last. Thus macrophages, which adhere very rapidly, will be found mainly in the first flask. Fibroblasts are the next most adhesive followed by epithelial cells. Haemopoetic cells such as T and B cells are the least adhesive and may be found, still in suspension, in the final flasks. In fact, macrophages will often migrate out of whole tissue fragments and attach to the substratum. The rest of the tissue can be lifted off and disaggregated for further separation.

Cloning by Selective Detachment

Some cell types are more easily detached from the substratum by the different mechanisms of action of enzymes such as trypsin and collagenase than are others. Embryonic fibroblasts are readily detached from the flask by a brief exposure to trypsin or collagenase whereas epithelial cells are more sensitive to treatment with the Ca^{2+} -chelator EDTA (Versene).

Physical Methods of Cell Separation

Cloning procedures are time consuming, so by the time a clone has produced enough cells to use, the line may well be close to senescence, particularly if the clones are derived from non-tumour tissues. Tumour tissue and continuous cell lines are cloned more rapidly but, because of the nature of these rapidly dividing cells, considerable heterogeneity can arise even as the clone is being grown up.

Where cells do not grow efficiently enough for cloning, or if time is limited it may be easier to employ physical separation techniques to produce populations of similar cell types. Cells may be separated on the basis of size, density (specific gravity), surface charge or surface chemistry (for example, by the ability to be bound by antibodies or lectins).

Separation Based on Cell Size

The relationship between cell size and sedimentation rate under gravity may be expressed as

$$v = \frac{r^2}{4}$$

where v = sedimentation rate (mm h^{-1}) and r = radius of the cell (mm).

Thus, in principle, cells may be separated on the basis of their size. The relationship shown above is, however, a gross oversimplification and many factors other than the radius of the cell may influence its sedimentation rate.

Cell density may also influence the rate of sedimentation. Obviously the properties of the suspending medium (e.g. viscosity, density) will also influence the rate of sedimentation. These may be used to improve the separation of different cell types.

Separation Based on Cell Density

For this, a density gradient is established in a centrifuge tube using a suitable density medium. Usually Percoll, Metrazamide or Ficoll is used for this purpose.

The cells are layered on the surface of the density gradient and centrifuged so that the cells sediment to a point where the density is equivalent to their own density. The cell layer can then be siphoned off with a pasteur pipette. This technique is known as *isopycnic sedimentation*.

Separation Based on Cell Surface Charge

With this technique the cells are separated by electrophoresis. The cell suspension is passed between two electrodes (polarized plates) and the cells migrate to either plate according to their net charge i.e. cells bearing a net negative charge migrate towards the positive plate and vice versa.

The real problem with this procedure is gravity. If the electrodes are placed vertically, all the cells will tend to sediment onto the bottom electrode. Thus:

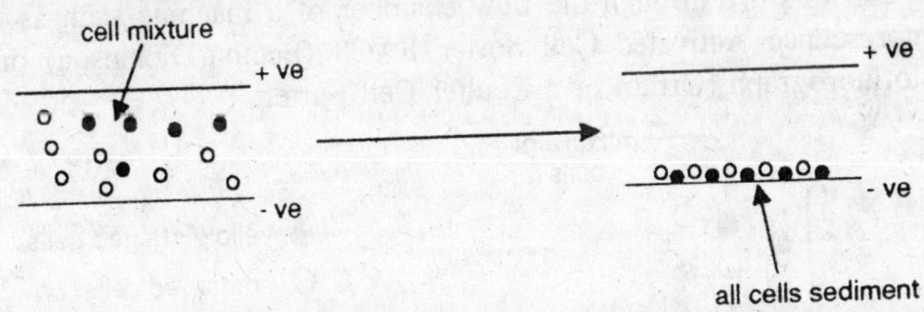

If we try to overcome this by using a higher voltage, there is a tendency for localized overheating and, hence, damage to the cells.

If the electrodes are held vertically, we will still tend to get cells settling out.

+ ve
- ve
+ ve
- ve
greater proportion of o
greater proportion of ●

However, some separation of cells can be achieved. (Note that urokinase producing cells were separated from a heterogenous kidney cell population using electrophoresis aboard an American satellite. The zero gravity enabled better separation of cells. This approach is not open to all!!)

A further difficulty is that populations of cells do not usually show qualitative differences in surface charge and we are usually faced with trying to separate cells which are all negatively charged. In these cases, we are dependent upon separating cells on the basis of their net charge:mass ratios. Despite these difficulties, electrophoresis is the basis of cell separation used in cytofluorometry.

Separation Based on Affinity

Cells in suspension can be passed through affinity columns which contain a matrix coated with antibodies or lectins. As the cells pass through they bind to or are captured by the matrix and can be specifically eluted by washing the column with detergent of enzyme solutions. Care has to be taken not to damage the cells during this process.

Separation by Cytofluorometry

Fluorescent dyes called fluorochromes can be used to label cell surfaces or cytoplasmic molecules. The cells are separated by passing the cell mixture through the flow chamber of a machine such as the Fluorescence Activated Cell Sorter FACS (Becton-Dickinson) or a cytofluorograph (Ortho) or a Coulter Cell Sorter.

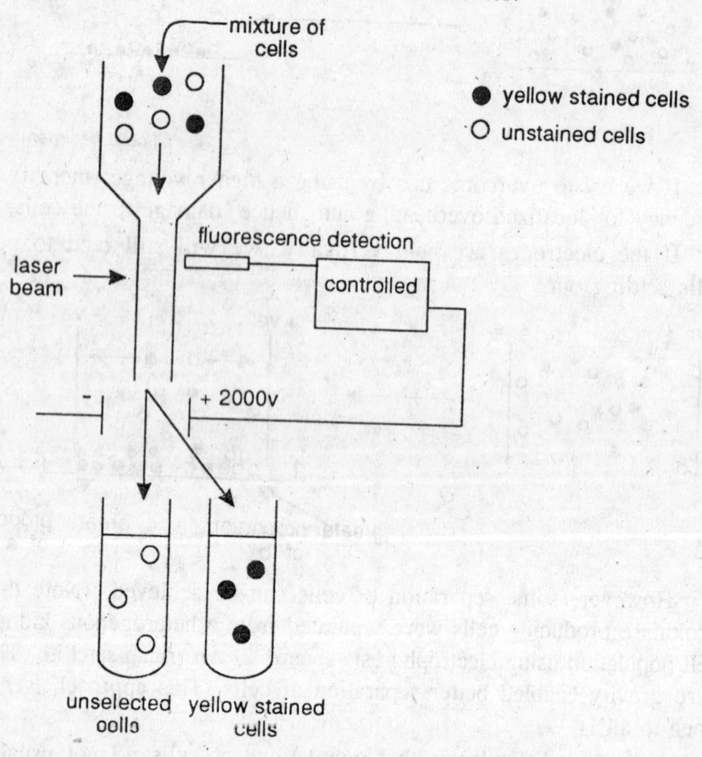

Fig. 3.5. A cell sorter based on cytofluorometry.

The fluorescent light produced by the cells under ultra violet light is detected by a photomultiplier and recorded. The cell sorter diverts cells which emit light within pre-set emission boundaries into a receiver tube placed below the cell strains. Those cells which do not emit within the pre-set emission boundaries are diverted to another tube. If one of these cells pass through the laser beam, it fluoresces and this is detected by the fluorescence detector. This, through the controller, switches on the voltage at the plates. The negatively charged cell is diverted into the right hand tube. After a short time (that is after the passage of the cell), the voltage is switched off.

If, on the other hand, an unstained cell passes into the laster beam, there is no fluorescence and the voltage is not switched on. Thus, such a cell is not diverted and passes straight through and enters the left hand tube.

This process is very fast and many cells can be separated in a matter of seconds.

This method can be used to separate cells on the basis of any differences that can be tagged by fluorescent labels, e.g. DNA, RNA, protein, enzyme activity or specific antigens or immunological markers on the cell surface such as HLA or viral antigens. The disadvantage of this method is that the cell yield is very limited and the equipment is very expensive both to buy and to run.

Separation by Flow Cytometry

This works in a similar way to separation by cytofluoremetry except that the light scattering potential of cells rather than fluorescent labels as used to distinguish between cell types.

Hazards and Safety in the Cell Culture Laboratory

Many of the reagents used in laboratories are hazardous if ingested or absorbed through the skin. Chemicals such as DMSO, dyes for staining cells and mutagens for transforming cells are often carcinogenic or mutagenic and must be handled with care. Gloves and laboratory coats must always be worn, masks should be worn when weighing out powdered dyes.

Biohazards, in the form of viruses, prions and oncogenes, must also be taken into consideration when working with animal tissue. Blood and blood products may contain Hepatitis viruses and HIV, brain and CNS tissue may carry prions which Call cause Creutzfield Jacob disease. Animal tissues, particularly from primates may contain Herpes B viruses which are lethal to humans though often carried asymptomatically by the natural hosts.

If tumour cells gain access to the body, they may continue to grow *in vivo* and produce a tumour in the new host.

4

STERILIZATION

All stocks of chemicals and glassware used in tissue culture should be reserved for that purpose alone. Traces of heavy metals or other toxic substances can be difficult to detect other than by a gradual deterioration of your cultures. It also follows that separate stocks imply separate glassware washing. The requirements of tissue culture washing are higher than for general glassware; a special detergent may be necessary and cross-contamination from chemical glassware must be avoided.

All apparatus and liquids which come in contact with cultures or other reagents must be sterile.

PROCEDURES FOR THE PREPARATION AND STERILIZATION OF APPARATUS

Glassware

Items of glassware used for dispensing and storage of media, and for cell culture, must be cleaned very carefully to avoid traces of toxic materials, contaminating the inner surfaces, becomes incorporated into the medium. Where the glass surface is to be used for cell propagation it must not only be clean but also carry the correct charge. Caustic alkaline detergents render the surface of the glass unsuitable for cell attachment and require subsequent neutralization with HCl or H_2SO_4, but many modern detergents do not alter the glass surface and can be removed completely.

For the most effective washing procedure: (1) Do not let soiled glassware dry out. A sterilizing agent, such as sodium hypochlorite, should be included in the water used to collect soiled glassware (a) to remove any potential biohazard and (b) to prevent microbial contamination, growing up in the water; (2) select a detergent which is effective in the water of your area, rinses off easily, and is nontoxic; (3) ensure that the glassware is thoroughly rinsed in tap water and deionized or distilled water, before drying; (4) dry inverted; and (5) sterilize by dry heat to minimize the risk of depositing toxic residues from steam sterilization.

Plastic culture flasks are, on the whole, meant for single use, as washing detergent renders them unsuitable for cell propagation (in monolayer) and resterilization is difficult. Cells may be reseeded back into the same flask after subculture but this tends to increase the risk of contamination.

Table 4.1. Sterilization of Equipment and Apparatus

Item	*Sterilization*
Apparatus containing glass & silicone tubing	Autoclave
Disposable tips for micropipettes	Autoclave
Dispenser tubing for Compupet	Autoclave
Filters—Millipore, Sartorius	Autoclave-do not prevac or postvac.
Glassware	Dry heat
Glass bottles with screw caps	Autoclave
Glass coverslip	Dry heat
Glass slides	Dry heat
Instruments	Dry heat
Magnetic stirrer bars	Autoclave
Pasteur pipettes—glass	Dry heat
Pipettes-glass	Dry heat
Screw caps	Dry heat
Silicone tubing	Autoclave
Stopper—rubber, silicone	Autoclave
Test tubes	Dry heat

Sterilization procedures are designed not just to kill the bulk of micro organisms but to eliminate spores which may be particularly resistant. Moist heat is more effective than dry heat but does carry a risk of leaving a residue. Dry heat is preferable but a minimum of 160°C for 1 hr. Moist heat (for fluids and perishable items) only be maintained at 121°C for 15-20 min. For moist heat to be effective, steam penetration must be assured and for this the sterilization chamber must be evacuated prior to steam injection. Insertion of Thermalog indicators monitors both temperature and humidity during sterilization.

Table 4.2. Sterilization of Liquids

Solution	*Sterilization*	*Storage*
Agar	Autoclave	Room Temperature
Amino acids	Filter	4°C
Antibiotics	Filter	–20°C
Bacto-peptone	Autoclave	Room Temperature
Bovine serum albumin	Filter-use stacked filters	4 °C
Carboxylmethyl cellulose	Steam–30min	4°C
Collagenase	Filter	–20°C
DMSO	Self-sterilizing, aliquot into sterile tubes	Room temperature kept dark, avoid contact with rubber or plastics
EDTA	Autoclave	Room temperature
Glucose–20%	Autoclave	Room temperature
Glucose–1-2%	Filter (low concentration caramelize if autoclaved)	Room temperature
Glutamine	Filter	–20°C
Glycerol	Autoclave	Room temperature
HEPES	Autoclave	Room temperature
HCL 1 N	Filter	Room temperature
Lactalbumin hydrolysate	Autoclave	Room temperature
Methocel	Autoclave	4°C
$NaHCO_3$	Filter	Room temperature
NaOH 1 N	Filter	Room temperature
Phenol red	Autoclave	Room temperature
Salt solutions (without glucose)	Autoclave	Room temperature
Serum	Filter-use stacked filters	–20°C
Sodium pyruvate 100 mM	Filter	–20°C
Transferrin	Filter	–20°C
Tryptose	Autoclave	Room temperature
Trypsin	Filter	–20°C
Vitamins	Filter	–20°C
Water	Autoclave	Room temperature

Materials

- Pipette cylinders (to collect used pipettes)
- Disinfectant (if required)
- Detergent

Soaking baths

Bottle brushes

Stainless steel baskets (to collect washed and rinsed glassware for drying)

Aluminium foil.

Sterility indicators (Browne's Tubes, Thermalog Indicators)

Glass petri dishes (for screw caps)

Autoclavable plastic film (Portex, Cedanco) or paper sterilization bags

Sterile-indicating autoclave tape

Collection and Washing

1. Collect immediately after use into detergent containing a disinfectant such as sodium hypochlorite. It is important that apparatus should not dry before soaking, or cleaning will, be much more difficult.
2. Soak overnight in detergent.
3. Machine wash or brush by hand or machine the following morning and rinse thoroughly in four complete changes of tap water followed by three changes of deionized water. If rinsing is done by hand, a sink spray is a useful accessory; otherwise bottles must be emptied and filled completely each time. Clipping bottles in a basket will help to speed up this stage.
4. After rinsing thoroughly, invert bottles, etc., in stainless steel wire baskets and dry upside down.
5. Cap with aluminum foil when cool and store.

Sterilization

1. Place in an oven with fan-circulated air at 160°C
2. Check that temperature has returned to 160°C, seal the oven a strip of tape with the time recorded on it, and leave for 1 hr.; ensure that the center of the load achieves 160°C by using a sterility indicator or recording thermometer with the sensor in the middle of the load. Do not pack the load too tightly; leave room for hot air circulation.
3. After 1 hr, switch off the oven and allow to cool with the door closed. It is convenient to put the oven on an automatic timer so that it can be left to switch off on its own.
4. Use within 24–48 hr. Alternatively, bottles may be loosely capped with screw caps autoclaved for 20 min at 120°C with prevacuum

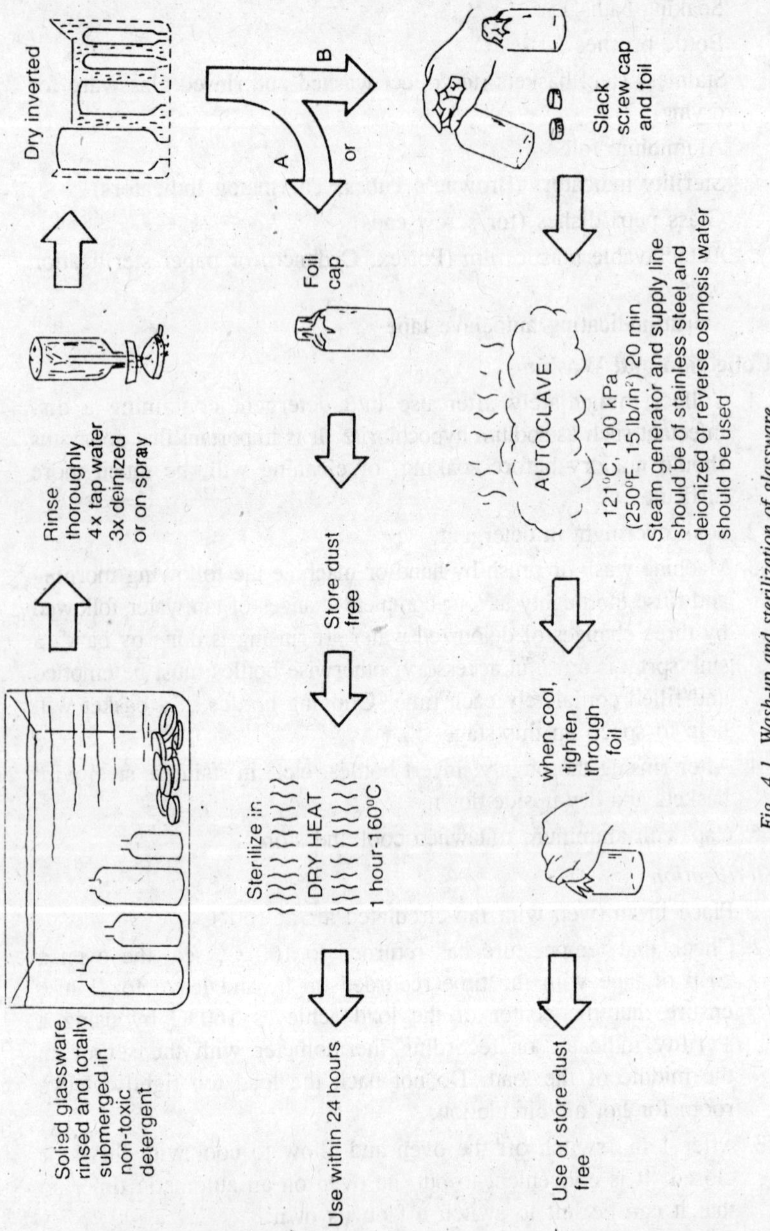

Fig. 4.1. Wash-up and sterilization of glassware.

and postvacuum cycle and the caps tightened when cool Caps must be very slack (one complete turn) during autoclaving to allow entry of steam and to prevent the liner being sucked out of the cap and sealing the bottle. If the bottle seals during sterilization in an autoclave, sterilization will not be complete. Unfortunately, misting often occurs when bottles are autoclaved and a slight residue may be left when this evaporates. There is also a risk of the bottles becoming contaminated as they cool by drawing in nonsterile air before they are sealed. Dry-heat sterilization is better, allowing the bottles to cool down within the oven before removal.

Pipettes

1. Place in water with detergent (e.g. 1% Decon) tip first and a sterilizing agent (hypochlorite or glutaraldehyde) immediately after use. Do not put pipettes which have been used with agar or silicones (water repellents, antifoams, etc.) in the same cylinder as regular pipettes. Use disposable pipettes for silicones and either rinse agar pipettes after use in hot tap water or use disposable pipettes.
2. Soak overnight and remove plugs with compressed air the following morning.
3. Transfer to pipette washer tips uppermost.
4. Rinse by siphoning action of pipette washer for a minimum of 4 hr in an automatic washing machine with a pipette adaptor.
5. Turn over valve to deionized water, or wait until last tap water finally runs out, turn off tap water, and empty and fill three times with deionized water (automatic deionized rinse cycle in a machine).
6. Transfer to pipette drier or drying oven and dry with tips uppermost.
7. Plug with cotton. Alternatively, pipette plugs may be dispensed with a short sterile filter tube placed on pipette bulb during use. To avoid cross-contamination, this must be replaced before starting work and every time that you change to a new cell strain.
8. Sort pipettes by size and store dust free.

Sterilization

1. Place pipettes in pipette cans (square aluminium or stainless steel with silicone cushions at either end; square cans do not roll on the bench) and label both ends of the cans. Put one pipette-size per can with a few cans containing an assortment of 1-ml, 2-ml, 10-ml, and 25-ml, pipettes in the ratio 1 : 1 : 3 : 2.
2. Seal with sterile-indicating tape and sterilize by dry heat for 1 hr at 160°C. This is different from the sterile indicating tape used

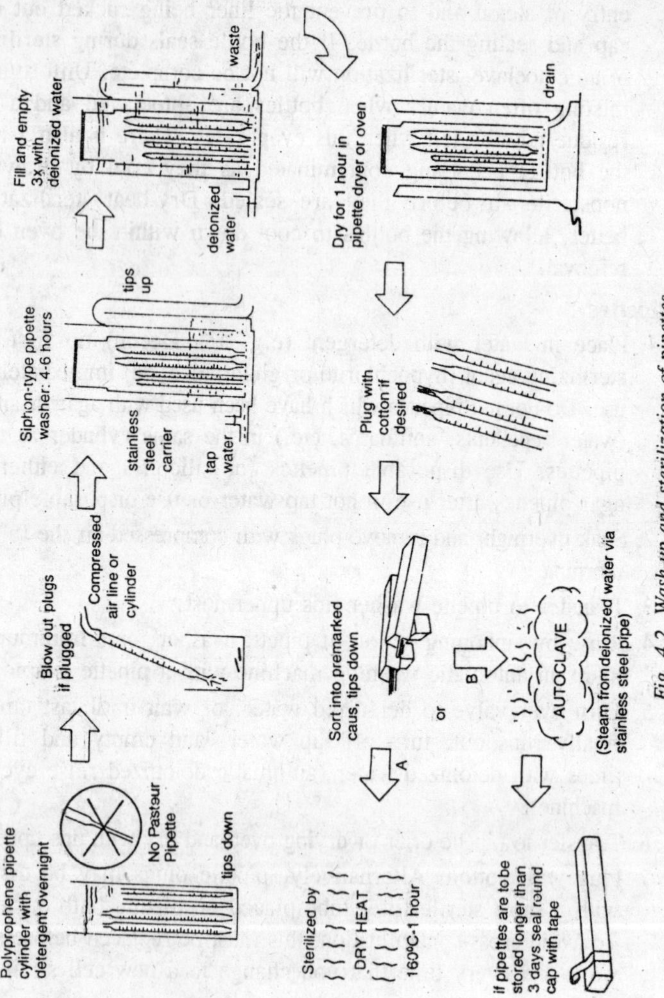

Fig. 4.2. Wash-up and sterilization of pipettes.

in autoclaves, as the sterilizing temperature is higher. Most tend to char and release traces of volatiles from the adhesive, which can leave a deposit on the oven, and potentially on the pipettes. Use the smallest amount of tape possible or replace with temperature indicator bars (Bennett), which are small and have less volatile materials. The temperature should be measured in the center of the load, to ensure that this, the most difficult part to reach, attains the minimum sterilizing conditions. Leave spaces between cans when loading the oven, to allow for circulation of hot air.

3. Remove from oven, allow to cool, and transfer cans to tissue culture laboratory. If you anticipate that pipettes will lie for more than 49 hr before use, seal cans around the cap with adhesive tape.

Screw Caps

There are two main types of caps for glass bottles in common use: (1) aluminum or phenolic plastic caps with synthetic rubber or silicone liners and (2) wadless polypropylene caps, reusable or disposable. The reusable polypropylene Duran caps are the best choice, as they are deeply shrouded and have ring inserts to improve the seal and improve pouring (although pouring is not recommended in sterile work). Disposable polypropylene caps will only seal if screwed down very tightly on a bottle with no chips or imperfections on the lip of the opening.

Do not leave aluminum caps, or any other aluminum items in alkaline detergents for more than 30 min, as they will corrode. Do not have glassware in the same detergent bath or the aluminum may contaminate the glass. Avoid machine washing detergents as they are very caustic.

Reusable caps

Soak 30 min in detergent and rinse thoroughly for 2 hr (make sure all caps are submerged). Liners should be removed and replaced after rinsing. Rinsing may be done in a beaker (or pail) with running tap water led by a tube to the bottom. Stir the caps by hand every 15 min. Alternatively, place in a basket, or better in a pipette washing attachment, and rinse in an automatic washing machine, but do not use detergent in machine.

Disposable caps

These should not need to be washed unless reused. They may be washed and rinsed by hand as above (extending the detergent soak if necessary). Because these caps may float, they must be weighted down during soaking and rinsing. For automatic washers, use pipette washing attachment and normal cycle with machine detergent.

Stoppers

Use silicone or heavy metal-free white rubber in preference to natural rubber. Wash and sterilize as for disposable caps. (There will be no problem with floating in washing and rinsing).

Sterilization

Place caps in a glass petri dish with the open side down. Wrap in cartridge paper or steam permeable nylon film (Portex), and seal with autoclave taps. Autoclave for 20 min at 121°C, 100 kPa (1 bar, 15 lb/in^2).

Keep organic matter out of the oven. Do not use paper tape or packaging unless you are sure that it will not release volatile products on heating. Such products will eventually build up on the inside of the oven, making it smell when hot, and some deposition may occur inside the glassware being sterilized.

Selection of Detergent

Solicit samples from local suppliers and test them: (1) for their ability to wash heavily soiled glassware; (2) for the quality of the growth surface afterwards; and (3) for the toxicity of the detergent.

Washing efficiency

Materials

Standard 75-cm^2 or 120-cm^2 glass culture vessels

Samples of detergents made up to working strength HBSS

Protocol

(i) Autoclave flasks carrying cell monolayers, standing vertically, for 20 min. at 120°C.

(ii) Soak overnight in detergent, three flasks per detergent.

(iii) Rise out detergent with water, not flasks with residue and brush as necessary to get them clean.

(iv) Rinse completely four times in water and three times in deionized water.

(v) Dry and sterilize by dry heat as above.

(vi) Check flasks again for apparent cleanliness and any sign of residue. Add a little BSS containing phenol red to each flask (approximately 1 ml/100 cm^2), rinse over the inside of the bottle, and look for any colour change. If it becomes pink (alkaline), detergent has not been completely rinsed out of the bottle. When you are satisfied that the flasks are clean, sterilize them by dry heat.

Quality of growth surface

Materials

Monolayer cells with good cloning efficiency (10% or more)

Medium

Serum

(for materials for fixing and staining cells).

Protocol

(i) Taking flasks from 1 (vi) (above), and celi suspension to each flask and to three control flasks (Disposable tissue culture grade

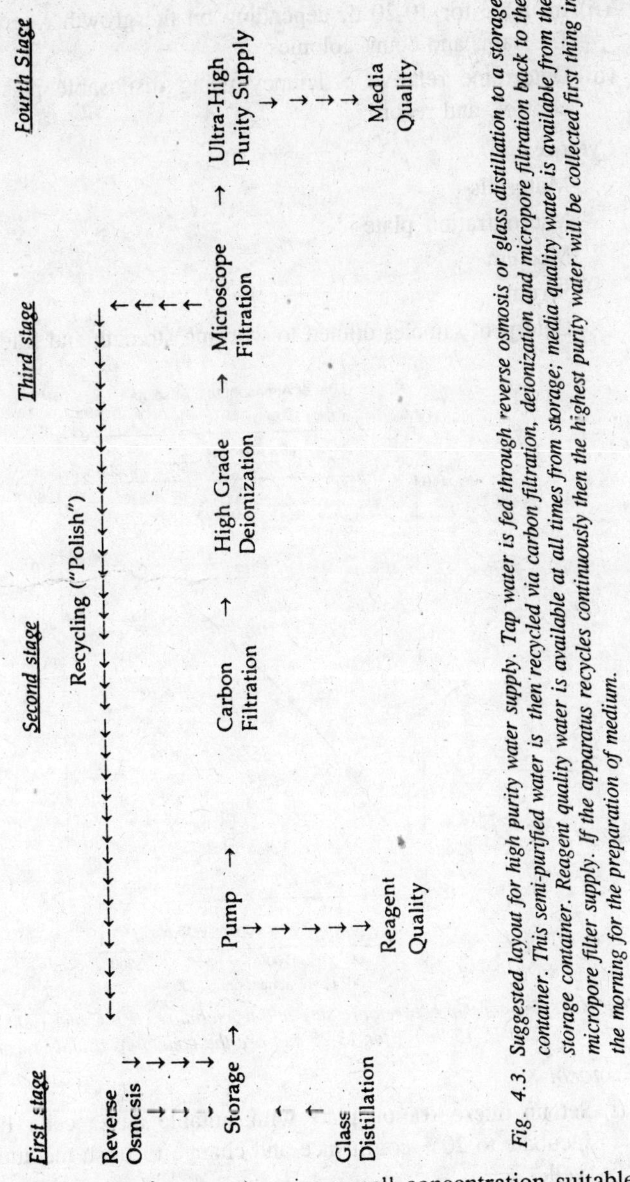

Fig 4.3. Suggested layout for high purity water supply. Tap water is fed through reverse osmosis or glass distillation to a storage container. This semi-purified water is then recycled via carbon filtration, deionization and micropore filtration back to the storage container. Reagent quality water is available at all times from storage; media quality water is available from the micropore filter supply. If the apparatus recycles continuously then the highest purity water will be collected first thing in the morning for the preparation of medium.

plastic of same surface area), using a cell concentration suitable for cloning (e.g., 20 cells/ml, 5 cells/ cm^2, CHO-K1; 100 cells/ ml 25 cells/cm^2, MRC-5. Use the minimum concentration of serum which will allow the cells to clone.

(ii) Incubate for 10-20 d, depending on the growth rate of the cells, fix stain, and count colonies.

(iii) Determine relative efficiency using disposable plastic flask as control, and record.

Cytotoxicity

Materials

Microtitration plate

Medium

Serum

Detergent samples diluted to working strength and filter sterilized.

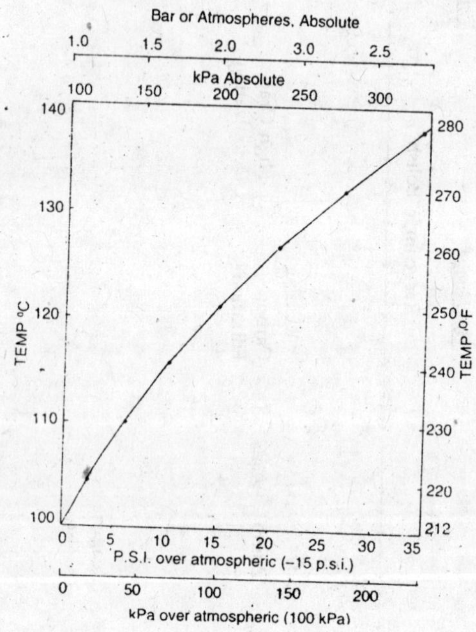

Fig. 4.4. Relationship between pressure and temperature: 121°C and 100 kPa, or 1 Bar (250°F, 15 lb/in²) for 15-20 min are the conditions usually recommended.

Protocol

(i) Set up microtitration plate with suitable target cells 1,000/well, incubate to 20% confluence and change to fresh medium, 100 μl/well.

(ii) Add 100 μl detergent diluted to usual working strength in medium and filter sterilized to the first well in each row and dilute serially across plate in that row, leaving the last two wells in each row as untouched controls.

(iii) Continue growth for 3-5 but *not beyond* the time that control wells reach confluence.
(iv) Wash, fix and stain plate.
(v) Determine titration point (point at which cell number per well is reduced by approximately 50%) and record. A high detergent concentration at this point means low cytotoxicity.

Miscellaneous Equipment

Cleaning

All new apparatus and materials (silicone tubing, filter holders, instruments etc.) should be soaked in detergent overnight, thoroughly rinsed, and dried. Anything which will corrode in the detergent—mild steel, aluminum, copper, or brass, etc.—should be washed directly by hand, without soaking, or soaking for 30 min only, using detergent if necessary, then rinsed and dried.

Used items should be rinsed in tap water and immersed in detergent immediately after use. Allow to soak overnight, rinse, and dry. Again, do not expose materials which might corrode to detergent for longer than 30 min. Aluminum centrifuge buckets and rotors must never be allowed to soak in detergent.

Particular care must be taken with items treated with silicone grease or silicone fluids. They must be treated separately and the silicone removed, if necessary, with carbon-tetrachloride. Silicones are very difficult to remove if allowed to spread to other apparatus, particularly glassware.

Packaging

Ideally, all apparatus for sterilization should be wrapped in a covering which will allow steam penetration but be impermeable to dust, microorganisms, and mites. Proprietary bags are available bearing sterile indicating marks which show up after sterilization. Semi-permeable transparent nylon film is sold in rolls of flat tube of different diameters and can be made up into bags with sterile-indicating tape. Although expensive, it can be reused several times before becoming brittle.

Tubes and orifices should be covered with tape and paper or nylon film before packaging, and needless or other sharp points should be shrouded with a glass test tube or other appropriate guard.

Sterilization

The type of sterilization used will depend on the material. Metallic items are best sterilized by dry heat. Silicone rubber (which should be used in preference to natural rubber). Teflon, polycarbonate, cellulose acetate, and cellulose nitrate filters etc., should be autoclaved for 20 min at 121°C, 100 kPa (1 bar, 15 lb/in^2) with pre-evacuation and post-evacuation steps in the cycle. In small bench-top autoclaves and pressure-cookers, make sure that he autoclave boils vigorously for 10-15 min before pressurizing to displace all the air. (Take care that enough water is put in at the start to allow for this). After sterilization, the steam is released and the items removed to dry off in an oven or rack.

Take care releasing steam and handling hot items to avoid burns. Wear elbow length insulated gloves and deep face well clear of escaping steam.

Sterilizing filters

Reusable filter holders should be made up and sterilized as follows:

1. After thorough washing in detergent rinse in water, then deionized water and dry.
2. Insert support grid in filter and place filter membrane on grid. If polycarbonate, apply wet to counteract static electricity.
3. Place prefilters (glass fiber and others as required) on top of filter.
4. Reassemble filter holder, but do not tighten up completely (leave about one half turn on collars, one whole turn on bolts).
5. Cover inlet and outlet of filter with aluminum foil.
6. Pack filter in sterilizing paper or steam-permeable nylon film and close with sterile indicating tape.
7. Autoclave at 121°C, 100 kPa (1 bar, 15 lb/in2) with no pre-evacuation or post evacuation ("liquids cycle" in automatic autoclaves).
8. Remove and allow to cool.
9. Do not tighten filter holder completely until the filter is wetted at the beginning of filtration.

Alternative Methods of Sterilization

Many plastics cannot be exposed to the temperature required for autoclaving or dry heat sterilization. To sterilize such items, immerse in 70% alcohol for 30 min and dry off under uv light in a laminar flow cabinet. Care must be taken with plexiglass (Perspex, Lucite) as it may crack in alcohol or uv treatment due to release of stresses built in during manufacture.

Ethylene oxide may be used to sterilize plastic, but 2-3 wk are required for the Et_2O to clear from the plastic surface.

γ-irradiation, 2,000-3,000 by (20-100 krad), is the best method for plastics. Items should be packaged and sealed. Polythene may be used and sealed by heat welding.

Reagents and Media

The ultimate objective in preparing reagents and media is to produce them in a pure form (1) to avoid the accidental inclusion of inhibitors and substances toxic to cell survival, growth, and expression of specialized functions, and (2) to enable the reagent to be totally defined and the functions of its constituents fully understood.

Most reagents or media can be sterilized either by autoclaving if they are heat stable (Water, salt solutions, amino acid hydrolysates) or by membrane filtration if heat labile. During autoclaving the container should be kept sealed if borosilicate glass or polycarbonate. Soda glass bottles are better left with the caps slack to minimize breakage. Evolution vapor will help to prevent ingress of steam from the autoclave, but the liquid level will need to be restored with sterile distilled water later.

Media and reagents supplied on line to large scale culture vessels and industrial or semi-industrial fermentors can be sterilized on line by ultra high temperature treatment for a short time (Alfa-Laval). Adaptation of this process to media production might allow increases automation and ultimately reduce costs.

Water

Water must be of a very purity for use in tissue culture, particularly if serum-free media are required. As water supplies vary greatly, the degree of purification that is required may vary. Hard water will need a conventional water softener on the supply before entering the purification system, but this will not be necessary with soft water.

There are four man approaches to water purification: distillation, deionization, carbon filtration, and ultrafiltration. Simple systems depend on the first two alone or combined, while the more efficient systems employ multiple stages, each operating on a different principle.

5

BASIC TECHNIQUES

Primary explanation denotes the cultivation of pieces of tissue fresh from the organism. The primary explanation techniques are the traditional techniques of tissue culture and, with very few exceptions, they were almost the only techniques in use until about fifteen years ago. These methods are still widely employed and will obviously continue to be for a very long time to come. Most of the techniques are similar in principle, the main differences being in the vessels used for growing the tissue. The main groups of methods are those using depression slides, Carrel flasks and roller tubes, of which there are many variations. Organ cultures, which logically fall into the same general category, actually require a highly specialized collection of techniques which separate them from the ordinary tissue culture methods. Methods employing the same principles as those which were developed for vertebrate tissues have also been employed for the cultivation of insect and plant tissues.

SLIDE CULTURES

Slide cultures or coverslip cultures are made by explanting a very small fragment of tissue on to a coverslip which is subsequently inverted over the cavity of a depression slide. This, the oldest and the traditional form of tissue culture, is still quite widely employed in some fields and still has its place. It has a number of advantages. Thus, it is simple and relatively inexpensive. Also, the cells spread out in a manner which is favourable for microscopic examination and photography in the living state. Finally, the cells grow directly on the coverslip so that they are easily fixed and stained and made into permanent preparations. The disadvantages of the technique are that the small supply of oxygen and nutrients is very rapidly exhausted so that the medium quickly becomes acid and this necessitates that rapidly grown tissues must be transferred frequently. In addition, it is difficult by this technique to maintain sterility for long periods and, finally, only very small amounts of tissue can be cultured, consequently the application of the method is limited. From the foregoing it will be appreciated that the main place of slide culture is in morphological studies. Modifications of the method are particularly valuable in time-lapse cinemicrogrphic investigations.

There are six general types of slide cultures:

1. Single coverslip with plasma clot (Harrison).
2. Double coverslip with plasma clot (maximow).
3. Single coverslip with fluid hanging drop (Lewi and Lewis).
4. Double coverslip with perforated cellophane (Schilling and Earle).
5. Thin drilled metal or glass slide with plasma and fluid medium (Gey).

The Preparation of Slide Culture

Single coverslip with plasma clot

This technique is still commonly employed and is probably the one that has been used most in tissue culture during the past fifty years. Details of the technique are as follows.

1. Set out all the instruments required, including a rack with sterile test-tube for holding Pasteur pipettes. Prepare the constituents of the medium. Usually this is made up in two parts, one part containing plasma and the other part containing embryo extract. Typically, one solution would contain 50 percent plasma is balanced salt solution and the other one would contain 50 per cent embryo extract in serum.
2. With a pair of sterile forceps, place on or two coverslips (22 mm) on a clean, sterile surface (a bench top which has been wiped with 70 per cent alcohol and allowed to dry). Use a capillary pipette to place one drop of the plasma-containing solution in the centre of each coverslip.
3. Transfer one or two explants to this drop, either by means of the knife blades or by a fine pair of forceps, taking care not to crush the tissue.
4. Add to this a drop of the embryo extract-containing solution. Immediately mix thoroughly before clotting starts and spread out

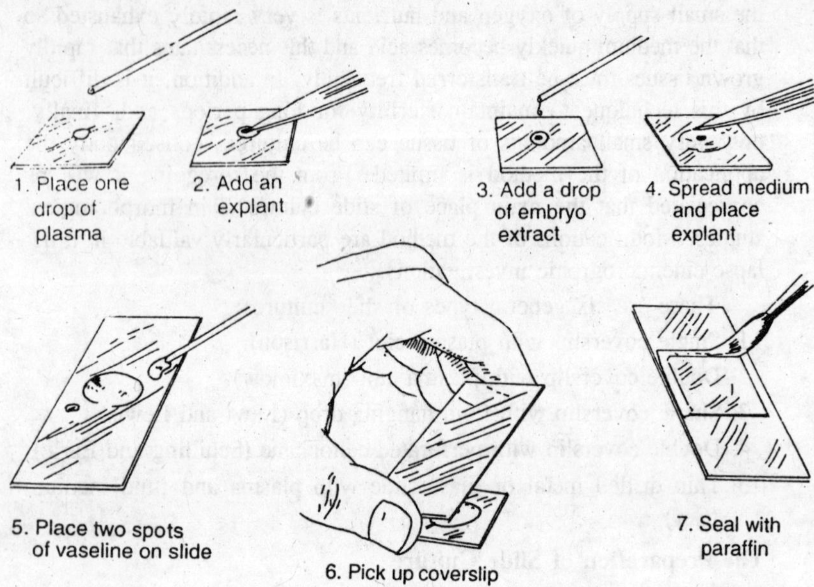

Fig. 5.1. Preparing coverslip culture.

into an area of approximately 15 to 20 mm. diameter. Locate the explants in this according to the desired arrangement.

5. With a glass rod, place two small spots of vaseline near the concavity of a depression slide in such a position that they will be covered by the coverslip. Invert the slide over the coverslip preparation and press down in such a way that the vaseline sticks the coverslip to the slide. Place the culture aside to permit the medium to clot.
6. Turn the cultures over and seal the margins of the coverslips with hot paraffin. Label and incubate at 37°C. Sealing the slide with paraffin must be done with care. It is better to use a rather thick seal since, apart from being more effective, this is easier to remove. The wax is conveniently kept in a double saucepan and should be thinned with about 25 or 30 per cent vaseline before use. A wax based on beewax is more satisfactory than one based on paraffin wax. Instead of wax a useful sealing material for short-term cultures is rubber solution (as used for repairing bicycle tyres).

Maximow double coverslip method with plasma clot

This technique is very similar to that already described. One or two large (40 mm) square coverslip are placed on a sterile surface and small drop of balanced salt solution is put on each. A square or round 22 mm. Coverslip is then placed in the centre of each large one, being held in position by surface tension of the balanced salt solution. The culture is then prepared on the small coverslip exactly as before. A large depression slide is used and the entire preparation is attached to it by vaseline and wax in such a way that the small coverslip is not in contact with the slide at any point.

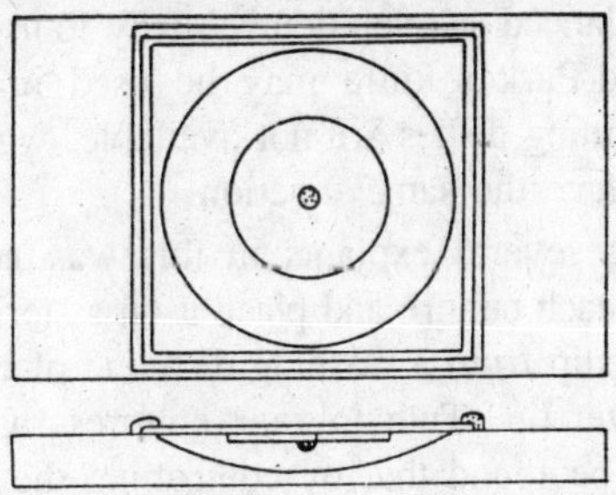

Fig. 5.2. Maximow double coverslip method.

Single coverslip with liquid Medium—lying and hanging drop cultures

This is an extremely simple technique but its applications are very limited. The prepared explants are placed in culture medium in a watch glass. Coverslips are laid out on a sterile surface, the explants drawn into the tip of a capillary pipette and deposited, one in the centre of each coverslip. Fluid can be drawn off and carefully measured (1 drop for a 1-inch slide, 2 to 3 drops for a large slide). The fluid must then be spread out in a very thin circular film with the explant protruding above its surface. A vaselined depression slide must be applied immediately and the preparation turned over with a quick flip to prevent the fluids from running into the crevice between the slide and coverslip. After labelling and paraffining, slides may be incubated in either the upright or inverted position. The cells grow directly on the glass of the coverslip.

After-Care of Slide Cultures

Single coverslip cultures are very useful for short-term studies but they are difficult to handle subsequently except by a process of transfer. Carrel's immortal strain was kept going for over 30 years by this technique but it is really extremely tedious and is not recommended. The Maximow double coverslip technique was developed to facilitate handling of long-term coverslip cultures where it was desirable to leave the explant in its original location.

Washing and Feeding Double Coverslip Cultures

1. By means of razor blade, remove the paraffin seal, then with the finger and thumb remove the large cover slip with small one attached. Flip it over so that the culture is uppermost and lay it on a clean surface.
2. With a needle in the left hand and forceps in the right, detach the small coverslip and transfer it to a Columbia staining dish containing balanced salt solution. Usually four coverslip, a pointed (Np.11) Bard-Parker knife may be used in the same way. If Columbia staining dishes are not available, watch glasses or Petri dishes will serve the same function.
3. After treating several explants in this way select a large clean coverslip for each culture and place it on a sterile surface. Remove a small coverslip from a washing dish and place it, culture up, on the large coverslip. Two to four cultures may be handled at a time. It will be found that by controlling the rate of withdrawal of the small coverslip from the balanced salt solution, the amount of fluid left adhering to it can be regulated fairly precisely. If too much remains, difficulties will be incurred subsequently in feeding the culture. If there is too little, air bubbles will appear between the two coverslips making observation difficult.
4. Feed the culture by adding a drop of feeding solution (e.g., 1 part of balanced salt solution, 1 part of serum, 1 part of 50 per cent embryo extract) to the small coverslip and attach a clean vaselined depression slide as before.

Patching

Sometimes it is necessary to patch the plasma clot if there is any evidence of liquefaction. The procedure is as follows. First, wash the cultures as before. Then in a watch glass place two drops of a mixture of plasma and balanced salt solution. To this add two drops of a mixture of serum and embryo extract. Mix these quickly and place a drop of the mixture on each coverslip before coagulation begins. The pipette and watch glass must be discarded after each patching. A clean vaseline slide may then be attached. *Note* — Never place a pipette contaminated with embryo extract in the tube of plasma.

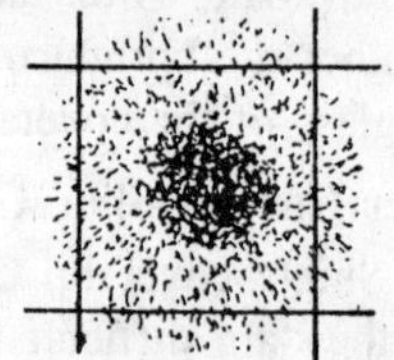

Fig. 5.3. Cutting an explant for transfer to a new clot.

Transferring Coverslip Cultures

Coverslip cultures have to be transferred by cutting away the excess plasma from the explant and then dissecting it and re-explanting each of the resulting pieces on a new coverslip. Place the coverslip on a raised block. With a Bard-Parker knife and a small curved blade make a cut through the outgrowth as illustrated. Note that uninfiltrated plasma must not be included since otherwise further growth from the explant will not occur. Lift the knife cleanly away after cutting so as not to tear the clot. The square of tissue may then be cut into two or four pieces and each of these transferred to new coverslips and treated as new explants.

In setting up slide cultures the main mistakes made by beginners are as follows:

1. Damage to the explant by excessive crushing or tearing during preparation.
2. Inadequate washing of the explant with the result that erythrocytes and debris cloud the clot.
3. Opacity of the clot from other causes such as bubbles, or dirty embryo extract.

Carrel Flask Technique

This technique is now used very much less than formerly but is still has its proponents and therefore merits a description. Its main use nowadays is in the establishment of strains from fresh explants of tissue, by the technique described by Earle et al. For many purposes, other types of flask are quite satisfactory, e.g. Erlenmeyer flasks. A good Carrel flask, however, has excellent optical properties and this is an advantage if it is desired to follow growth of the cells microscopically. The design of the Carrel flask is of importance since those available commercially are very variable. It is particularly important that the neck should be as wide as possible since a narrow neck makes manipulations almost impossible. It is also desirable to have a flask which has excellent optical properties but very few of these are available since the manufacture of an optically-perfect *Carrel flask* is a highly skilled operation.

The main advantages over coverslip cultures are as follows: This tissue can be maintained in the same flask for months or even years. Large numbers of cultures can be prepared fairly easily and fairly large amounts of tissues can be grown. Large amounts of medium can be used and this makes it possible for the medium to be removed for

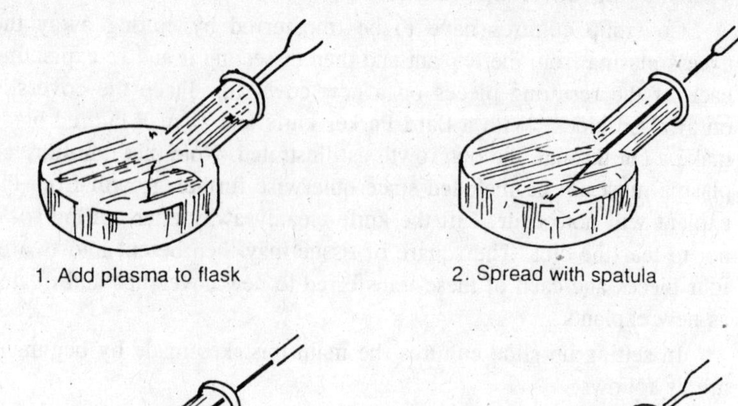

Fig. 5.4. Preparing Carrel flask culture.

assay. In flask cultures the gaseous phase can be readily controlled and amount of medium can be measured accurately.

There are two types of Carrel flask techniques, thick and thin clot cultures. The thick clot encourages rapid growth and is particularly suitable for short-term cultures or for the maintenance of strains. The clots can be removed intact from the flask for staining. On the other hand, thin clot cultures are best for maintaining cultures over a considerable period of time and they are better suited for testing the effects of materials added to the medium.

Preparation of Cultures

1. Place some D 3.5 (diamter of 3.5). Carrel flasks in a rack with their necks pointing to the right. A convenient number to handle is up to six at any one time. The necks should be flamed.
2. Place one drop of plasma on the floor of each flask. Using a platinum spatula spread this plasma out in a circle extending just short of the edge of the floor of the flask.
3. With the spatula, transfer the desired number of explants to the plasma and allow clotting to occur. The explants can be arranged according to a desired pattern by manipulation with the spatula.

Dr. Parshley uses a convenient planting guide which is placed under the flask and which shows where the explants should be situated.

4. After the plasma has clotted and the explants are fixed in position, extra medium can be added. In the case of thick clot cultures 1.2 ml. of dilute (1/4) plasma are placed on top of the explants and the whole is then left to clot. For thin clot cultures, 1.2 ml. of dilute (1/4) serum is added instead of the plasma.
5. Finally, the flasks should be gassed with an appropriate gas phase, usually 5 per cent of carbon dioxide in air.

Renewal of Medium

1. The old fluid is drawn off by means of a pipette.
2. 1.2 ml. fluid medium is added to replace this.
3. The flask is again gassed.

In the case of thick clot cultures, the flask should be inverted and allowed to drain for several hours before replacing the used plasma. In either case, if there is any liquefaction of the clot a patch should be made by adding a drop of fresh plasma and allowing it to clot.

The Transfer of Tissue

If it is desired to transfer the tissue from these flasks to other vessels it has to be removed and cut into suitable pieces before being replanted. In the case of thin clot cultures the explant can be scraped off with a spatula and removed for cutting. On the other hand, in dealing with thick clot cultures the entire clot must be carefully loosened with a spatula and slid out entire on to a glass plate before cutting up the explants.

A common use of the Carrel flask technique, as has been mentioned previously, is the establishment of cell strains. For this purpose, a strip of tissue 5 to 10 mm. long and 1 to 2 mm. wide is planted in a thick clot. After it has begun to grow well the tissue is removed and replanted. It is a common practice to trypsinize the entire clot after the culture has become established and to use the suspension to seed new vessels. In this way, the cells can be transferred from a plasma substrate to a glass substrate. The procedure to be used for trypsinization.

TEST-TUBE CULTURES

Ordinary test-tubes from very cheap and convenient vessels for the cultivation of cells and tissues. They can be used for preparing large numbers of cultures and can be set up either in roller drums or in stationary racks. They have the advantage of being cheap and easy to handle in large quantities. There are, however, a number of disadvantages in the use of test-tube cultures. The optical conditions are very poor. Also it is very difficult to quantitate accurately due to the curvature of the inside of the tube. There is a definite risk of contamination of ordinary stoppered tubes due to a slight leak of air or medium between the stoppers and the tubes. This may particularly occur on opening tubes which have been removed from the incubator when there is almost invariably an inrush of air. This is likely to carry with it any organisms which may have settled between the stopper and the lip of the tube. If a small drop of medium has been left there during a previous transfer the region is almost certain to be heavily contaminated. This complication must always be kept in mind in handling test-tube cultures.

The actual techniques of test-tube culture resemble closely those employed with other vessels. Cultures may be grown in plasma clots, in which case the technique is very similar to the Carrel flask technique. They may also be grown directly on the glass wall of the vessel without a plasma clot and this technique requires a little more consideration. Lastly, cell suspensions may be allowed to settle and grow on the inside of test-tubes and this technique is exactly the same as that employed with other types of vessels.

Plasma Clot Technique

1. With a Pasteur pipette, place a drop of plasma near the bottom of each tube and spread it over the lower third of the tube.
2. Transfer the explants, usually about four, to the plasma and leave them to clot.
3. After they are fixed in position, add medium, usually 0.5 to 1 ml. per tube.
4. Stopper and label the tubes and place them in a stationary rack or in a roller drum in the incubator.

Feeding Test-Tube Cultures

Feeding test-tube cultures is simply achieved by removing the supernatant fluid and replacing it with fresh medium.

Patching Test-Tube Cultures

If there is evidence of liquefaction of the plasma clot then, after removal of the supernatant fluid, one drop of plasma mixed with one drop of culture medium should be added to the tube and it should be rotated slightly to ensure that the culture area is covered. After the plasma has clotted, fresh fluid medium may be added.

Transfer of Cultures from Test-Tubes

Cultures can very readily be removed from test-tubes by means of a bent Pasteur pipette. With the tip a small circular cut is made in the plasma surrounding the chosen colony. By gently pushing on the margins of the dissected plasma the small disc containing the colony is freed from the glass vessel. It is then transferred to a glass plate and trimmed to remove the peripheral plasma in the same way as for coverslip cultures. Note that in all procedures involving the opening of test tubes which contain cultures, great care should be taken to flame the mouth of the tube very thoroughly indeed. If the stopper has become wet by contact with medium, it is best to discard it and use a new one.

Culture of Primary Explants in Roller Tubes without Plasma

It is possible to grow fragments of tissue direct on glass without the use of a coagulum, and this technique has been used quite commonly with roller tubes. In order to prevent the explants from falling off the wall of the tube, particularly when it is being washed by fluid continuously in a roller drum, it is necessary to allow them to adhere by partial drying. The fragments of tissue are taken up into a Pasteur pipette along with some medium and are transferred to the tube. This is rolled around gently as the liquid containing the explants is added. The fragments may thus be left adhering to the tubes and after they have all been placed the excess medium is removed. The tubes are then stoppered and left to dry for 10 to 15 minutes 0.5 to 1 ml. of medium is added and the tubes are placed on a rack or in a roller drum. The cells grow directly on the glass wall of the tube. A similar technique to the one described has been used in biochemical studies with large numbers of explants.

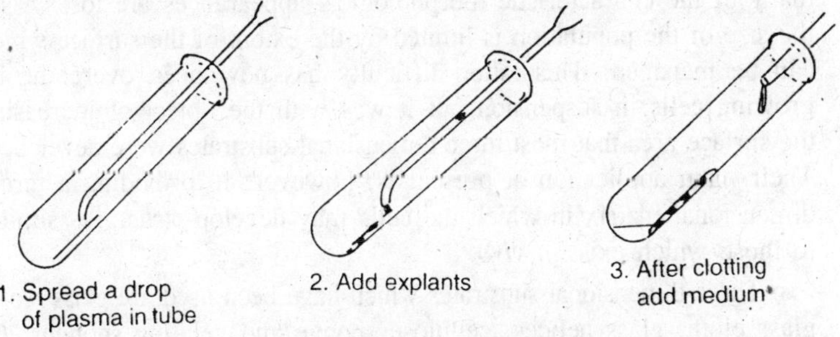

Fig. 5.5. Preparing roller-tube cultures.

Flying Coverslips in Test-Tubes

A very convenient way of way of preparing coverslip cultures for histological examinations or for mounting in perfusion chambers is by means of so-called *flying coverslips* in test-tubes. Flying coverslips are small narrow coverslips (11 mm. wide, 40 or 22 mm. long). These can be inserted into ordinary 6" × 5/8" test-tubes. Cultures are prepared on them in the standard manner for coverslip cultures but instead of mounting them on depression slides, they are slid into test-tubes. A fluid medium is added to the tubes (1 to 2 ml. each) which may then be placed in a stationary rack or in a roller drum. Feeding is simply performed by removing the fluid medium and replacing it with fresh material. Large numbers of cultures can be conveniently prepared in this way.

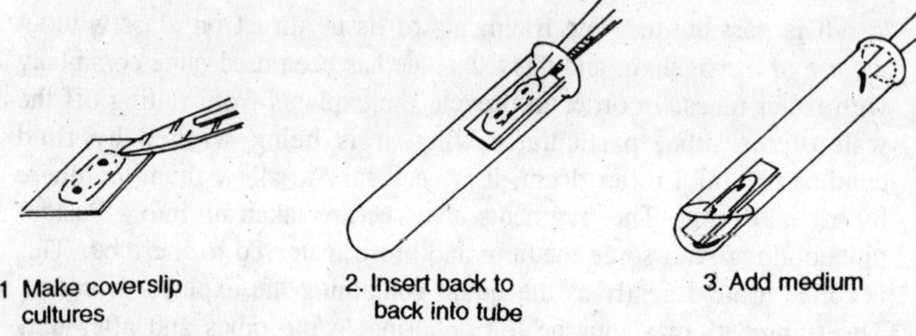

Fig. 5.6. Preparing flying coverslip cultures in roller tubes.

Three-Dimensional Substrates

In those techniques which have been described, the cells are allowed to migrate and grow in an essentially two-dimensional environment. This produces very flat cells spread over a large area. As a result the internal structures are beautifully displayed but, on the other hand, many of the characteristic morphological appearances are lost. Also, the size of the population is limited by the extent of the surface which can be managed. This latter difficulty has now been overcome by growing cells in suspension but it was with the object of increasing the surface area that most three dimensional substrates were developed. Their main application at present is, however, in providing a three-dimensional matrix in which the cells may develop structures similar to those which exist *in vivo*.

Three-dimensional substrates which have been used are glass wool, glass cloth, glass helices, cellulose sponge and gelating sponges. Of these, cellulose sponge, particularly in combination with plasma, is the only one used to any extent.

The material used by Leighton is fine-pore cellulose sponge, manufactured by Dupont. This is used for removing excessive moisture from photographic plates and is available from Ilford Ltd. It is prepared by first cutting it into strips of about 5 × 10 mm. cross-section. These are then held between glass slides and sliced with a razor blade. Slices about 0.5 mm. thick should be used and any that happen to be thicker should be rejected. They are prepared by boiling for total of one hour in two changes of distilled water and then washed for 30 minutes each in acetone, ether and absolute alcohol. After a final boiling in distilled water the slices can be autoclaved and are ready for use.

Cultures are prepared according to any of the standard procedures already described, the cellulose sponge being incorporated in a plasma clot along with the explant. The migrating cells invade the interstices of the sponge. These cultures are not very satisfactory for direct examination and are usually fixed and sectioned first. Leighton recommends fixation in Zenker-formalin for 30 minutes, followed by washing in water for two hours. The sponges can then be stored as required in 80 per cent alcohol. Before sectioning they are taken through alcohols and toluol (15 minutes each change) to paraffin and sectioned at 6 microns.

Organ Culture

Although this technique has been used to grow intact embryonic organs, the name is actually something of a misnomer since it is usually employed to maintain small pieces of organs *in vitro*. The object of the organ culture technique is to maintain the architecture of the tissue and to direct it towards normal development such as occur *in vivo*. In order to achieve this arm, it is essential that the tissue should never be disrupted or damaged and this requires careful handling, so that organ culture techniques generally demand more careful manipulation than tissue culture technique.

Media used for growing organ cultures are generally the same as those used for other types of tissue culture. Nowadays, synthetic or semi-media are very commonly used and, on the whole, have proved very satisfactory. The traditional medium is a thick plasma clot, containing a high concentration of embryo extract, the preparation of which requires special care. Twelve to fourteen day embryos are used. After removal of the gall-bladder, the remainder is pulped by cutting repeatedly with scissors in BSS. It is allowed to stand for 60 minutes and then centrifuged. The supernatant constitutes the required extract.

The technique of organ culture can be divided into those employing a solid medium and those employing a fluid medium.

The classical technique of organ culture consists of placing a small, carefully dissected piece of tissue on top of a plasma clot. The plasma clot is formed in a watch glass which itself rests on a pad of cotton wool in a Petri dish. The cotton wool is kept moist to prevent excessive evaporation from the dish. The tissue fragment invariably gives rise to some liquefaction of the plasma in its immediate vicinity and has to be transferred to a new situation every day or two. This requires some care in handling. The technique is rarely employed in the above form nowadays since it has been improved by using a raft of lens paper or rayon net on which the tissue is places. Transfer of the tissue can then be achieved easily by moving the raft very slightly. In the techniques most commonly employed today, the raft is used but the plasma clot has now been dispensed with and replaced by a liquid medium on which the raft floats. The use of a raft of lens paper was first described by Chen and the paper he used was supplied by G. Gurr & Company. Only a few types of lens paper are satisfactory since others rapidly become saturated with medium and sink. This difficulty has been overcome by Shaffer who employs, instead of lens paper, a raft of rayon acetate treated with silicone.

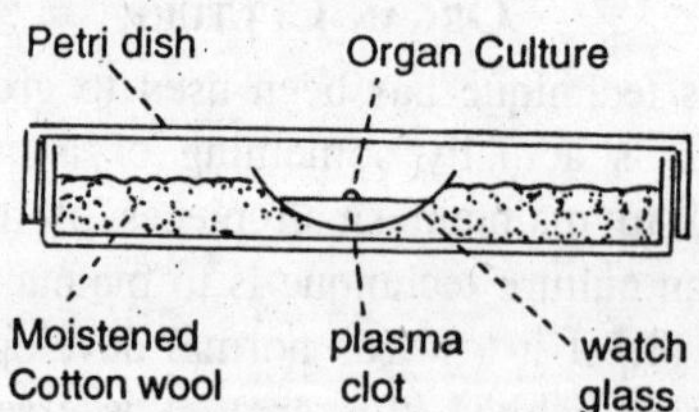

Fig. 5.7. The 'Classical' organ culture technique.

Preparing an Organ Culture on a Cellulose Acetate Raft (Shaffer)

1. A suitable cellulose acetate fabric with holes 0.51-mm square is selected and cut into pieces of suitable size. Shaffer used voile. Silene V.V. & Co. but this is no longer obtainable. The essential property of a suitable rayon is solubility in acetone. The pieces are prepared by three successive washes in glass-distilled water, two in absolute alcohol and two in ether, over a period of some hours. They are immersed in a silicone solution, e.g. Siliclad (Clay-Adams) 1 part in 100 parts water or MS 1109, drained, allowed to dry and left in a damp atmosphere overnight. They are washed with several changes of glass-distilled water, rinsed with

absolute alcohol, cut into pieces of suitable size (2 × 1 cm.) and sterilized by dry heat at 130°C (no to be exceeded).

2. The raft is placed on a very small drop of balanced salt solution (only enough to fill the interstices of the net) and the tissue is placed on it. The net is lifted from the drop of saline and excess fluid is removed by touching it against dry glass momentarily. It is gently placed in the surface of 0.5 ml. of medium in a watch glass.
3. Feeding of the culture is performed by removing the net, with its culture, removing excess fluid as before and placing it on top of a fresh pool of medium.

Note– If the siliconed fabric is wet wetted with a proteinaceous solution it will no longer float so that care must be taken in handling to prevent it this.

In addition to the use of rayon raft other worker have used special apparatus. For instance, Trowell has used a tantalum gauze platform of such a size that when medium is placed in the bottom part of the vessel it just comes level with the platform. The material to be cultures is then placed on it.

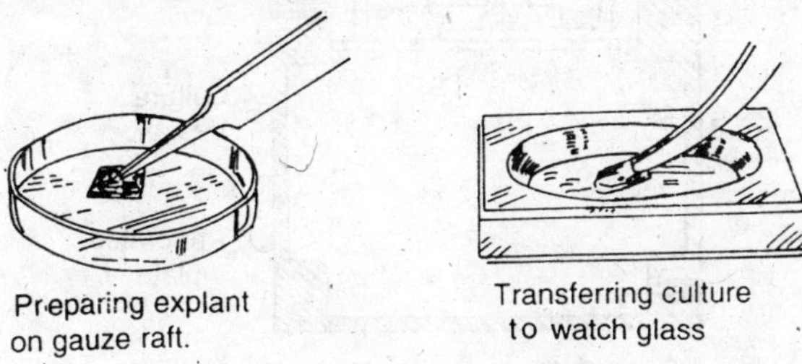

Fig. 5.8. Preparing an organ culture on a rayon raft.

Yet another organ culture technique involves the use of an agar substrate in which the essential nutrients for the tissue are dissolved. This technique has been used by Professor and Madame Wolff, particularly for the growth of embryonic organs. The essential nutrients in the medium are combined with a warm agar solution such that the final concentration of agar is 1.5 per cent. The agar-containing solution is well mixed (at about 45°C) and then left aside at room temperature to set. The organ are cultured on the surface of the solid substrate.

The technique so far described have been applied almost entirely to the cultivation of embryonic organs or embryonic organ fragments. The cultivation of fragments of adult tissues as organ cultures has proved a particularly difficult problem and Trowell has adduced evidence that is because of the much greater requirement of adult tissues for oxygen. Using the special medium T8 and special apparatus permitting the use of 95 per cent. O_2 in the gas phase, Trowell has successfully cultured a variety of adult organs, including liver. In these studies he found that serum was toxic and it should be noted that his experiments were done with serum-free media.

Trowell's Type II culture chamber consists of an aluminum chamber which can be sealed by a glass plate and which can be gassed with a mixture of 95 percent. O_2 and 5 per cent CO_2. The chamber is divided into upper and lower chambers by a perforated plate. The upper chamber contains a culture dish and the lower chamber serves as a gas reservoir. Within the culture dish is a square metal grid made of stainless steel expanded metal. This is made by taking a strip 25 × 33 mm and bending it about 4 mm from the ends to form a grid 25 mm square, standing on 4 mm legs.

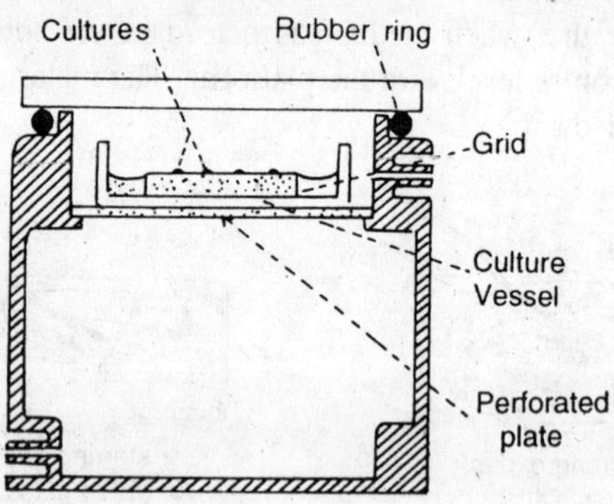

Fig. 5.9. Trowell's type II culture chamber.

The culture vessel is filled with medium so that it just comes up to the top of the grid and a piece of *lens paper* 27 mm square is placed on top so that it just becomes wet. The organ fragments are planted on this. Squares of agar gel may be substituted for lens paper if so desired. The tissue fragments should not exceed 2 mm. in diameter: otherwise central necrosis occurs.

In setting the apparatus it should first be gassed at 37°. After placing the organ fragments in position the apparatus should be gassed again before incubating. The tissues may be removed for sectioning after 6-9 days of cultivation. The essential feature of the organ culture technique is the prevention of migration of cells from the tissue fragments. This is achieved either by moving the tissue fragments frequently or by dissecting the tissue with great care so that no damage is done such as would promote migration, or by using a substrate such as agar or *rayon-acetate* on which cells will not normally migrate.

A kind of organ culture can be achieved by growing fragments of tissue in suspension. This technique has been in use for many years and, as the Maitland culture, is well known in virology. A similar technique has also been used by Parker for immunological studies and by Paul and Pearson for biochemical studies.

Chopped Tissue Technique

In the techniques described so far, each fragment of tissue is individually manipulated and planted. While this generally provides the best circumstances for controlled growth of an intact tissue, it severely limits the amount of tissue that can be handle and consequently for the large scale cultivation of viruses or for biochemical investigations, these techniques have a restricted application. To overcome this, a number of methods have been developed.

Trypsinization of fresh tissues to obtain uniform cell suspensions and the use of strains are elegant methods of handling large numbers of cells but they are limited to a few tissues. A somewhat cruder technique employing suspensions of minced tissue has been in use for many years for the cultivation of large amounts of fresh tissue, both for virus growth and biochemical investigations. It is of particular importance since it has been widely used for growing viruses in the manufacture of vaccines. With some refinements it can be used quantitatively with a fair measure of reproductibility.

The technique was first used by Maitland and Maitland in 1928 for the cultivation of vaccinia virus in suspensions of minced kidney in a mixture of serum and *Tyrode's solution*. Parker used it for a study of antibody formation in spleen, and an essentially similar technique has been used by Chen and ourselves for morphological studies on the one hand and biochemical studies on the other.

Cultivation of Poliomyelitis Virus in Minced Tissue Suspensions

This represents at once the simplest and most important application of the method to date. Fresh monkey kidneys are decapsulated and halved. The pelvis of the kidney is removed, leaving mainly the outer cortex. This is cut into small pieces and then chopped rather finely in Morgan and Parker's medium 199. The chopped tissue is transferred to 1 litre Roux flasks, along with some medium 199. The flasks are placed in a rocker which keeps the tissue agitated by rocking it back and forth through an angle of about 45° about twenty times a minute. After 24 hours the culture is inoculated with virus. The material is ready for harvesting in another four to seven days and then processed for vaccine manufacture.

It is desired to obtain more quantitative results, the following procedure is recommended, using the McIlwain tissue chopper to cut uniform explants.

Cutting Chick Embryonic Heart Explants by means of McIlwain Tissue Chopper

1. Remove the metal cutting table and sterilize it by dry heat. Also sterilize a number of razor blades for the instrument and a supply of filter-papers (9 cm). Swab the cutting arm, the moving platform and all other accessible part with 80 per cent ethanol and allow to dry.
2. Remove the hearts from chick embryos of up to 15 days. With a pair of scalpels cut off the atria and large vessels. Discard these.
3. Split the remaining ventricles in their greatest plane to give two halves from each heart. Wash in BSS.
4. Replace the metal cutting table (handling in the paper in which it was sterilized to avoid contamination). By means of forceps, place a razor blade in holder and adjust it, screwing the clamp for the blade very tight. Place a sheet of filter-paper on the platform and moisten with about 0.5 ml. BSS. Start the machine. Add other filter-papers and readjust the blade if necessary until it is cutting evenly into the top filter-paper but not through it.
5. Place the halved hearts on the cutting table with the inside down and the rather tough pericardium up. Press them down into the moist filter-paper with the forceps used to handle them.
6. Having set the chopper to cut slices of 0.25 to 0.75 mm, set the machine in operation. When all the tissue has been sliced turn the cutting table at right angles to the previous direction. Re-set the platform and repeat the cutting in this direction.
7. Remove the diced tissue by means of a sterile platinum wire with the end flattened to form a spatula. Transfer this tissue to a container with growth medium and, by rotating the wire rapidly, ensure that the fragments are separated from each other.
8. The fragments may be pipetted by means of a wide-mouthed pipette or handled individually, according to the purpose for which they are to be used.

WHOLE EMBRYO CULTURE

It will be remembered that probably the first recorded successful explanation was performed by Wilhelm Roux in 1885 when he maintained the medullary plate of a chick embryo for some time in saline. The technique was developed by Spratt for testing the effect of metabolic inhibitors on embryonic development. In the form he developed it constitutes a fairly straight forward technique which however, requires

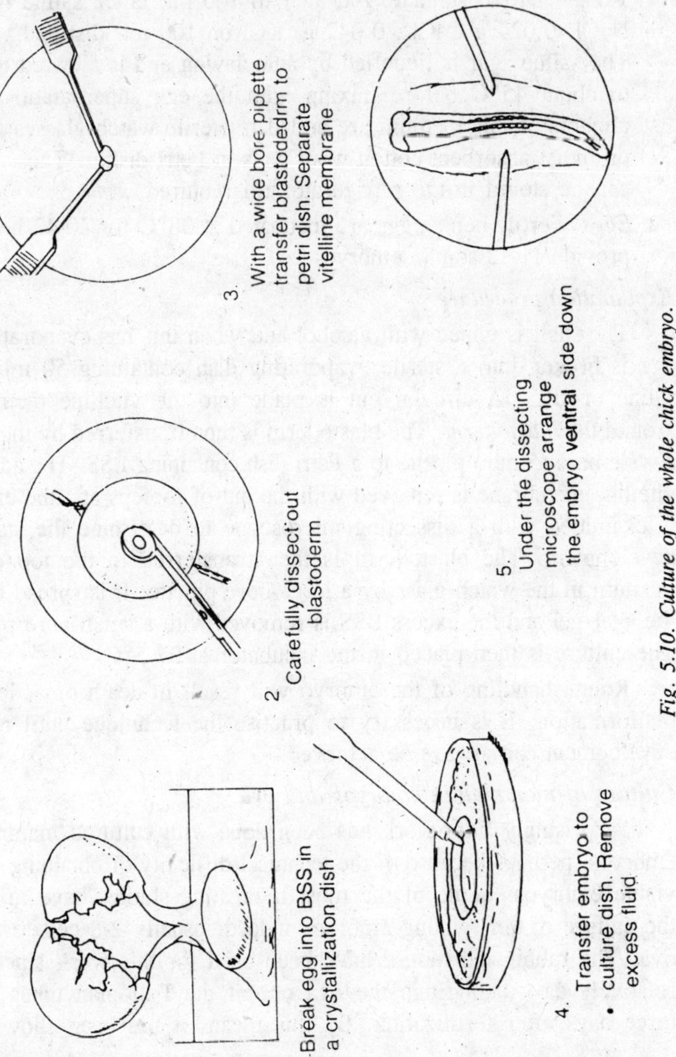

Fig. 5.10. Culture of the whole chick embryo.

considerable care in manipulation. 40 hour embryos are used and development can be followed for a further 24-48 hours *in vitro* before the embryo dies. The method is as follows.

Materials

1. *Medium.* Rothfel's modification of Spratt's medium is prepared by thoroughly mixing the supernatant from a centrifuged whole egg with an equal amount of saline-agar, which is prepared by adding 1.5 g. 'Difco' standardized agar to 100 ml. chick saline (0.7 g. NaCl; 0.024 g $CaCl_2$; 0.042 g. KCl on 100 ml. distilled water). The saline-agar is liquefied by autoclaving and is allowed to cool to about 45°C before mixing with the egg supernatant. 1 ml. aliquots of this medium are added to sterile watch glasses placed on moist absorbent cotton-wool pads in Petri dishes. These plates can be stored in the refrigerator till required.
2. *Eggs*. Fertile hen's eggs are incubated at 38°C for 40-42 hours to provide 11-13 somite embryos.

Explanation procedure

The shell is wiped with alcohol and when this has evaporated the egg is broken into a sterile evaporating dish containing 50 ml chick saline or BSS. A circular cut is made into the vitelline membrane around the *blastoderm*. The blastoderm is then transferred by means of a wide-bore (5 mm) pipette to a Petri dish containing BSS. The adherent vitelline membrane is removed with the aid of forceps and the embryo is examined with a dissecting microscope to determine the stage of development. The blastoderm is then transferred to the top of the medium in the watch-glass by a large-bore pipette. It is spread out on the agar gel and the excess BSS is removed with a small-bore pipette. The culture is then placed in the incubator at 37.5°C.

Rough handling of the embryo will result in death or at least in malformation. It is necessary to practise the technique until normal development can always be achieved.

Culture of mammalian embryos and ova

Surprisingly little work has been done with cultured mammalian embryos, perhaps because of the technical difficulty of obtaining young viable embryos. Some of the most interesting studies have involved the culture of very young embryos indeed, usually 2-8 cell fertilized ova. The rabbit and mouse have been used for this work since it is relatively easy to syringe the ova out of the Fallopian tubes up to three days after fertilization. By this means it has been shown that considerable development can take place *in vitro* and, for instance, 8-cell mouse ova have developed to blastocysts. It has been shown that ova cultured in this way may be reimplanted and can go on to form healthy animals. The media used have generally been of a rather simple kind and have varied from 100 per cent, serum through the common types of plasma-embryo extract media to simple *Krebs-Ringer solutions* supplemented with 1 per cent, of thin egg-white or bovine albumin.

TECHNIQUES FOR THE CULTIVATION OF TISSUES FROM COLD-BLOODED ANIMALS AND PLANTS

Cultivation of Tissues from Cold-blooded Animals

The techniques of growing these tissues are identical with those for warm-blooded animals. The one important detain in which they differ is the incubation temperature. In many cases it is important to ensure that this does not rise too high and for many fish tissues it should be kept below 18°C.

A very beautiful demonstration of the migratory activity of epithelial cells can be seen in explain obtained from the tail of a fish, particularly one of the fan-tailed varieties of goldfish. The tail is disinfected with a mild disinfectant (70 per cent alcohol) and washed with sterile BSS before snipping off a piece. This can be explained in a chick plasma clot in a medium containing 10 per cent, chick embryo extract. If the temperature is kept at about 18°C considerable migration is already visible in four or five hours.

Amphibian tissues can be treated in a very similar manner and the usual incubation temperature is 26°C.

Cultivation of Plant Tissues

The techniques of plant tissue culture are similar to some of those used for animal tissues, the main difference being in the media. A brief description of two classical types of culture will be given to illustrate the principles.

Culture of Tomato roots

Sterile tomato roots are first obtained. The seeds are removed aseptically from a ripe tomato (the skin can be thoroughly washed and then swabbed with alcohol). They are placed on moist sterile filter-paper in sterile Petri dishes and left in the dark for a few days to germinate. Healthy roots, 2-3 cm. long, are then cut off with scissors and transferred to 125 ml. conical flasks containing 50 ml. of *White's plant medium*. The flasks should not be stoppered but merely covered with aluminum foil. The roots grow and can be transferred weekly by cutting off about 1 cm. from the tip of each and placing in new medium. White has maintained a strain of tomato roots for over twenty years by this method.

Culture of Carrot Callus

In this case the culture is grown on a solid substrate. A number of 6" × 1" test-tubes are prepared by inoculating with 10 ml. of White medium each and adding a pad of four thickness filter paper. A large, healthy carrot is cleaned thoroughly and dried on the outside. It is then broken to expose a sterile surface and with a sterile corkborer a core about 5 mm. in diameter is removed. This is transferred to a Petri dish and sliced into discs. One or two discs are placed on a filter-paper and the tube, capped with aluminum foil, is placed in a rack at an angle.

Some tissues are more demanding than others in their nutritional requirements and the preparation of some explains requires more technical skill but the principles of plant tissue culture are the same in most techniques. Solid substrates made by incorporating agar in the medium are commonly used, the explants being grown on the surface.

6

Special Techniques

Sources of Tissues

Tissues for culturing are usually obtained from two fundamentally different sources, embryos on the one hand and adults on the other. The distinction is made between these two because they require somewhat different treatment. Embryonic tissues are sterile from the beginning and have merely to be handed with aseptic technique in order to obtain usable material. They also tend to grow easily, migrate rapidly and undergo mitosis very soon after explanation. On the whole, fibroblasts tend to grow out from them more rapidly than other types of cells. Adult material, on the other hand, has often to be specially treated to ensure its sterility before culturing. Many adult tissues grow *in vitro* with difficulty and there is often a latent period of several days before any outgrowth becomes apparent. There is less tendency for fibroblasts to grow out from adult tissues and very often other cell types, in particular epithelial cells, grow rather well from adult material. Tumour tissues from adult animals behave in many ways like normal adult tissue, but, on the whole they tend to grow more readily and to grow more rapidly in the first few days. A further difference between embryonic and adult material is the fact that embryonic materials are usually capable of growing in completely *heterologous media*, whereas adult tissues often grow better in media of homologous origin.

Embryonic Tissues

Embryonic tissues may, of course, be obtained from any source. The most readily available are those from the chick, the mouse and the human. Tissues from all these grow well although those from avian sources tend to cease to grow after several generations whereas tissue of mammalian origin frequently become permanently established.

The classical source of tissue for culture studies is the chick embryo. It is particularly convenient in as much as it provides a sterile source of materials, readily accessible when required. Since it is very widely used, the dissection of the chick embryo will be described in detail.

Dissection of the Chick Embryo

1. Select an egg which has been incubated for 7-14 days (for general purposes 10-12 days is a good age to provide tissues).
2. Check that the egg contains a healthy embryo by candling. (Hold the egg in front of an aperture in a piece of cardboard in front of a bright light. The shadow of the embryo should be readily discerned, with a clear airspace at the blunt end of the egg. The blood vessels should be easily distinguished but there should not be sharp line of demarcation at the margin of the vascular zone).
3. Wipe the eggshell with alcohol and allow it to dry. Crack the shell over the airspace and remove it, taking care not to allow pieces to fall on the membrane below. With a pain of sterile forceps lift up the opaque white membrane to expose the embryo. Lift the embryo out very gently with a pair of curved forceps placed under the neck and place it on a Petri dish.
4. Lay the embryo on its back and with an open pair of forceps in the left hand spread out the wings. Insert a curved pair of forceps under the abnormal skin and rip it open up to the sternum. Insert the forceps just under the sternum and pluck out the heart. Place it in a drop of BSS in a Petri dish.
5. The *liver* can be recognized as reddish-purple, soft object. The curved forceps are carefully closed behind it and it is then plucked out also.
6. The remainder of the abdominal viscera, mainly crop, stomach and intestines are plucked out and transferred to a Petri dish.
7. Work the forceps round behind each *eyeball* and pluck them out of their sockets. Place one blade of a pair of forceps under the ridge of the eyebrow, close the forceps and pull out the *frontal bone*.
8. Cut off the head. Commencing near the tail of the embryo apply two pairs of (preferably wide and blunt) forceps. Works these forward, one after the other, so as to squeeze out the *spinal cord*—rather like toothpaste from a tube. The same manoeurve can be applied to the head.

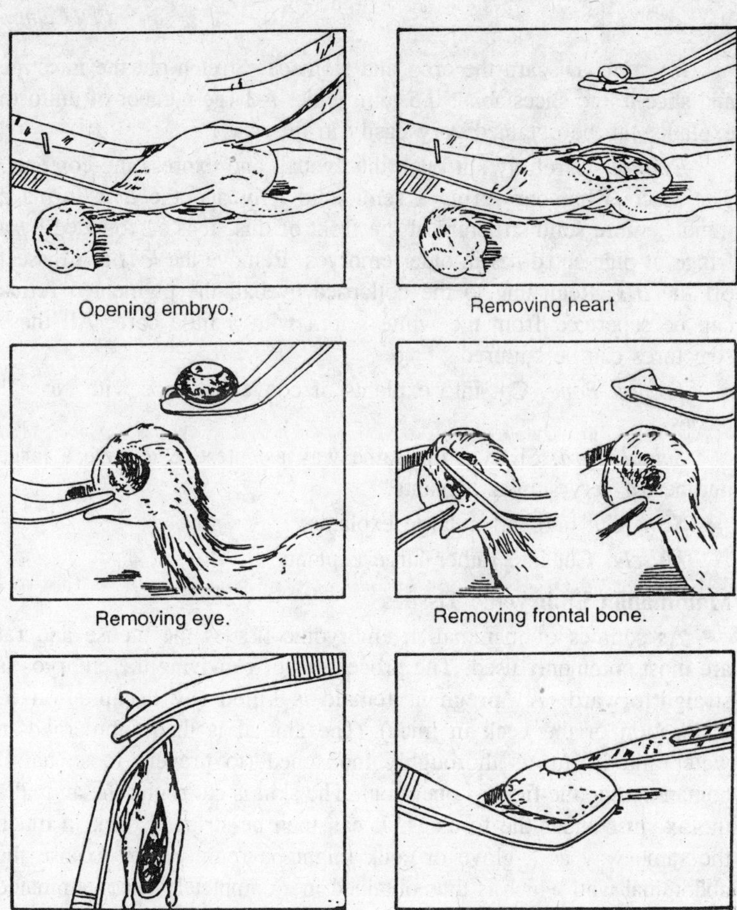

Fig. 6.1. Dissection of the chick embryo.

9. Skin is gently peeled from the top of the head, the back or the thighs.
10. *Skeletal muscle* is obtained from the thighs.
11. The various organs are subsequently treated as follows before explanation:

Heart. With two knives, fitted with No. 11 Bard-Parker blades, cut off the atria and great vessels. Wash the ventricles in BSS and cut up into explants of suitable size (0.5 mm. cubed).

Liver. Remove the gall-bladder (a small olive-green sac) and cut up the remainder into explants after washing with BSS.

Intestine. Discard the crop and stomach. Stretch out the intestine and slice it into slices about 0.5 mm. wide. A large number of uniform explants can be obtained very easily in this way.

Eyeball. Carefully slit open the eyeball and express the contents. The viscous humour forms a semi-solid jelly and the *lens* forms a minute, more solid structure at the front of this. It is surrounded by a fringe of pigmented *iris* in older embryos. Remove the *lens* and dissect off the *iris*. Returning to the collapsed eyeball the pigmented retina can be separated from the white sclera with a little care. All these structures can be cultured.

Frontal Bone. Cut into explants of convenient size with No. 11 blades.

Spinal Cord. Slice in the same way as intestine to give a large number of very similar explants.

Skin. Cut into rather large explants.

Muscle. Cut into rather large explants.

Mammalian Embryonic Tissues

As sources of mammalian embryonic tissues the mouse and rat are most commonly used. The procedure for removing the embryos is straightforward. A pregnant female is killed (by decapitation or dislocation of the neck in mice). The animal is then submerged in water and its fur is thoroughly moistened (to prevent loose hairs) contaminating the field of operator). The skin is cut round the animal's thorax, just under the forelegs. It can then be stripped back in much the same way as a glove or sock might be removed, to expose the abdominal wall which is thus obtained in a completely uncontaminated condition. It is easy then to open up the abdomen and remove the uterine horns containing the embryos. Care must be taken, of course, not to puncture the intestine. The uterus, placed in a sterile Petri dish, is opened and the embryos removed. Dissection of the mouse embryo can be conducted in a manner similar to dissection of the fowl embryo. It is, however, smaller and more skill is required.

Embryos of large mammals can be obtained from the slaughter-house and both pig and cow embryos provide excellent culture material. It is best to obtain the entire uterus containing the embryo. This is slung up from a hook and the outer surface of the uterus liberally swabbed with tincture of iodine (it is almost certainly very heavily contaminated). With a trocar and cannula most of the *amniotic fluid* is removed (to be used as supplementary medium if desired). The uterus is then placed in a large, clean tray, swabbed with iodine once more

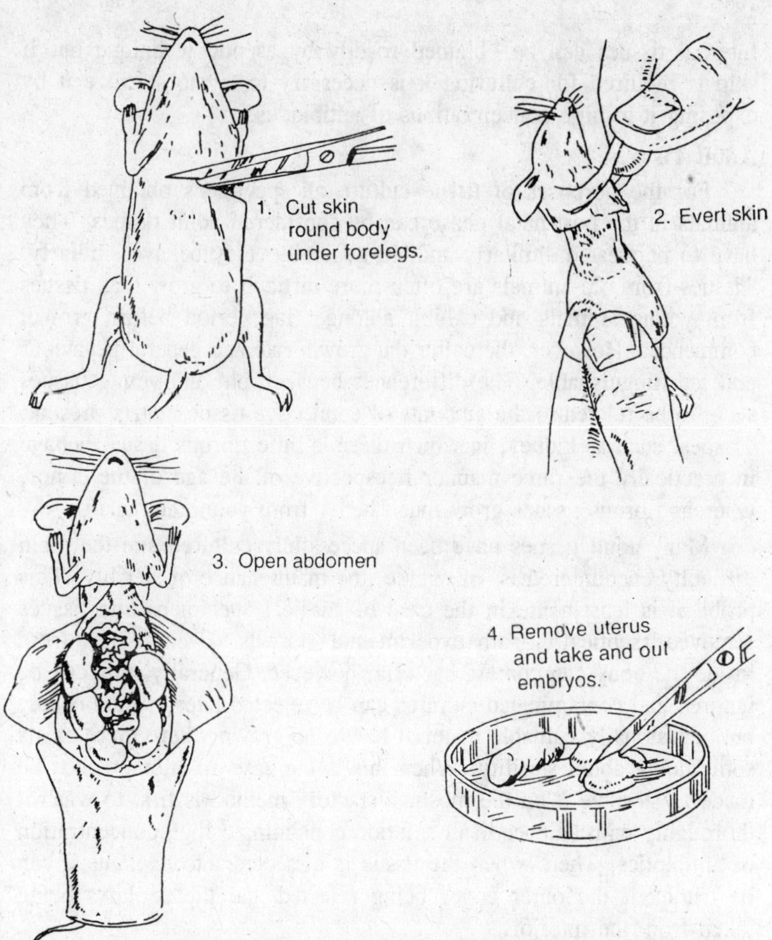

Fig. 6.2. Aseptic removal of embryos from a pregnant mouse.

and opened by a large incision. It is best to have an assistant retract the walls of the uterus as soon as the incision is made. The embryo can then be removed and transferred to a sterile container.

The human embryo also provides excellent material for tissue culture studies and embryos are readily available in some hospitals. Occasionally a *hysterectomy* specimen will become available and in that case it can be treated in exactly the same way as pig and cow uteri. More usually the embryo or foetus is likely to be obtained some time after death, almost certainly contaminated and possibly macerated. In this event it should be treated exactly as any other surgical specimen.

Internal tissues can be obtained readily by aseptic technique but if skin is required for culturing it is necessary to try to sterilize it by exposing it to high concentrations of antibiotics.

Adult Tissues

For the purposes of tissue culture all specimens obtained from animals in the post-natal phase can be considered adult tissues. They have to be treated similarly and, in generally, they behave similarly. Tissues from old animals are often more difficult to grow than tissues from young animals and exhibit a longer lag period before growth commences. However, thereafter the growth rate and general behaviour and indistinguishable. The differences between old and young tissues seem to be related to the amounts of connective tissue matrix present. Tissues, such as kidney, in which there is little fibrous tissue, behave in practically the same manner irrespective of the age of the donor, whereas fibrous tissues grow much better from young animals.

Many adult tissues have been successfully cultured and the main difficulty encountered is, of course, the maintenance of sterility. This problem is least acute in the case of surgical specimens and tissues removed aseptically from experimental animals. Even tissues from inside the body can contain bacteria, however. Generally, this can be ignored and contaminated cultures can be rejected later as they occur, but occasionally valuable material has to be grown where there exists some doubt about sterility. When this is the case an attempt must be made to sterilize it an the most-satisfactory method is first to wash it thoroughly with balanced salt solution containing a high concentration of antibiotics. Then, when the tissue is dissected into explants it can be trimmed, the outer edges being rejected and the explants being taken from the interior.

Among the classical adult tissues for cultivation is the central nervous system of the kitten which provides very suitable material for some morphological studies and has been intensively used for this purpose. The subcutaneous fibroblast of the adult rat is another tissue that has been commonly employed, particularly in studies of mitosis. Bone marrow from many animals, especially rabbits, has been particularly used in studies of *haemopoiesis*. Human surgical specimens, in particular lymphatic tissue, nasal epithelium, bone marrow and tumours, have provided material for many studies also.

Two adult tissues are of particular importance in virology—the kidney and the amnion. Kidneys are obtained usually from monkeys, by an aseptic operation. Human amniotic membranes are obtained from normal births, the midwifery staff being asked to collect them in sterile basins with the minimum of handling and the minimum of exposure to disinfectants. As soon as possible after delivery, the membranes are dissected free of the placenta and umbilical cord and are washed in sterile BSS containing *streptomycin* and *penicillin*.

One other tissue, the buffy coat of the blood, requires to be considered separately. The buffy coat is of particular interest since it provides a sample of cells which can be repeatedly obtained from the same individual, its behaviour is particularly reproducible and the phase of development of these cultures has been thoroughly described. The migration and growth of cells from the buffy coat of many animals has been examined but those most commonly used have been the fowl, the rabbit and the human. The procedure is as follows.

Preparation of Explants of the Buffy Coat

1. Coat several centrifuge tubes, test-tubes, Pasteur pipettes and serum needles with silicone (Repelcote, Hopkins and Williams). Similarly, prepare one or two 20 or 50 ml. syringe, depending on the amount of fluid required. Sterilize all the glassware, Place the centrifuge tubes and test-tubes in an ice-bath.
2. Remove the blood by venepuncture (or cardiac puncture in small animals) and immediately place 10 ml. in each of the centrifuge tubes, avoiding frothing. Seal the centrifuge tubes with a sterile cotton-wool plug turned down over the edge and secured with a rubber band, or with aluminium foil, and centrifuge at 1,800 g. for 10 minutes (preferably in a refrigerated centrifuge at 2-4°).
3. By means of a siliconed *Pasteur pipette*, transfer almost all the clear supernatant layer of plasma to a siliconed test-tube. Store this is an ice-bath or in the refrigerator.
4. To the surface of the packed cells remaining in the centrifuge tube and one drop of 1/10 dilution of embryo extract. Stir the surface very carefully with the end of the pipette and immediately centrifuge at 1,800 g. for another two or three minutes.
5. Handling them carefully so as not to disturb the cell layers, place the centrifugé tubes in an incubator. The surface will have clotted firmly in 10-15 minutes. With a sterile platinum wire manoeuvre the clot out of the tube into a Petri dish containing about 10 ml. of BSS. Wash it twice more in 10 ml. amounts of BSS to remove most of the red blood cells and then place it on a Petri dish with a few drops of BSS and cut it into explants in the usual manner.

6. These explants develop best in plasma clot and they can be implanted in the autologous plasma which has been stored in the ice-bath. A suitable clot is formed with 25 per cent plasma in Eagle's medium with a very small amount of dilute (!/10) embryo extract added to initiate coagulation.

Note—It is sometimes difficult to prevent premature coagulation of blood, particularly if it has been collected by an unpracticed operator by cardiac puncture. Coagulation can be further inhibited in this case by adding to each of the centrifuge tubes 0.1 ml. of 25 mg. per cent heparin before adding the blood. If this is done it is necessary to use a higher concentration of embryo extract to coagulate the plasma and undiluted EE_{50} should be used instead of 1/10 dilution. Although perfectly satisfactory results can be obtained by this technique, it is better to avoid if it possible since both heparin and embryo extract are harmful to cultures of peripheral blood cells and the best results are definitely obtained when anticoagulants are avoided.

Culture of Peripheral Blood Leukocytes

Although numerous reports concerning the cultivation of leukocytes from the peripheral blood by the above technique have appeared in the literature the results, especially with some mammals including man, have been somewhat erratic. The use of *phytohaemagglutinin* (originally in order to facilitate separation of erythrocytes from leukocytes) seems to stimulate a burst of mitosis in these cells and hence provides a very convenient way of obtaining samples of human cells in mitosis. These are particularly useful in the study of chromosomal abnormalities, such as *Mongolism* and *Klinefelter's syndrome*. The following method is adapted from that of Hungerford et al. which is itself a modification of the method proposed by Osgood.

1. To a volume of aseptically collected heparinized venous blood add phytohaemagglutinin in the proportion of 0.2 ml. per 10 ml. blood. Leave to stand in ice water for 30-60 minutes.
2. Centrifuge at 350 r.p.m. for 10 minutes at 4°C. Pipette off the supernatant containing the leukocytes.
3. Centrifuge at 2,000 r.p.m. for 3-4 minutes, remove the supernatant and resuspend the cells in a medium consisting of 20 per cent of the donor's plasma plus 80 per cent medium 199, containing antibiotics. The initial cell concentration should be about 1-2 $\times$ 10^6 cells per ml.
4. Inoculate test-tubes with about 2 ml. each of the cell suspension and incubate at an angle at 37.5°C.

5. After 3-4 days the mitotic index is at its maximum. If it is desired to a accumulate mitoses colchicine should be added to give a final concentration of 10^{-7} M and the cells should be harvested 8-18 hours later and prepared by a chromosome-spreading technique.

Storage of Tissue before Culturing

As a general principle it is best to set up cultures as soon as practicable after the tissue has been obtained. However, it is possible to use tissues which have been stored up to two days before use. Cells may be grown from foetuses which have been dead for 24 hours but post-mortem material is usually unsatisfactory unless it is removed almost immediately after death. The main factors affecting the viability of tissue are the temperature at which it is stored and the ease with which nutrients and oxygen can diffuse to it. Thus tissue in the centre of a large organ will soon die, particularly if the temperature remains fairly high, since metabolites in the vicinity of each cell will rapidly be used up when the circulation stops and no more metabolites will become available. On the other hand peripheral tissues, such as the skin, may survive for a very long time since they cool down quickly (so that metabolism is reduced) and oxygen can diffuse in from the air while metabolites can diffuse out from dying tissues in the deeper layers.

If precautions are taken to ensure that metabolism is reduced and metabolites can diffuse to the tissue survival for many days may be obtained. These conditions are achieved by cutting the tissue into small fragments, putting them into a nutrient solution and storing at room or refrigerator temperature. This procedure should be adopted if tissue cannot be handled immediately for any reason.

Having emphasized the importance of explaining the tissue as soon as possible after removal from the animal it should, however, be added that there is rarely any noticeable diminution in the growth potential of any tissue within the first two to four hours after removal so that there is plenty of time to obtain specimens and set them up.

Tissues from Cold-Blooded Animals and Insects

The problems met with in culturing these tissues are very similar to those met with in growing contaminated surgical specimens. The main problem is achieving sterility. When internal tissues from fairly large animals are required it is, of course, a simple matter to sterilize the outside with iodine or merthiolate and remove the internal tissues aseptically. Where this is not possible, washing in BSS containing high concentrations of antibiotics is the method of choice.

Tissues from Plants

The principles involved in obtaining plant tissues for culturing are exactly analogous to those used in obtaining animals tissues, *i.e.*, aseptic removal or disinfection.

Aseptic removal of tissue is fairly easy with many plant tissues. For instance, it is only necessary to break a carrot to expose a sterile surface from which a piece of tissue can be removed by means of a sterile cork-borer to give an excellent subject for cultivation. Similarly, the *potato*, *artichoke*, etc., can be sterilized on the outside with disinfectant and a slice removed to expose a sterile surface.

Twigs, branches, tree-trunks, and so on can be treated similarly. After sterilization of the outside the epithelium or bark can be removed to expose the sterile internal tissues, ready for cultivation.

In addition to tissues of the above type a good deal of plant tissue culture has been concerned with the growth of root tips. They are best obtained by removing seeds from the interior of a fleshy plant and permitting them to germinate in aseptic conditions. As a rule, the seeds are sterilized with detergents and hypochlorite, or some similar means, and then placed on moist sterile filter-paper or agar in the dark to germinate.

Preparation of Cell Suspensions from Fresh Tissues

Rous and Jones, 1916, described the use of trypsin to digest away the plasma clot in which a tissue culture was grown and showed that the cells could be replated on coverslips after they had been dispersed in this way. At that time the method seemed to have little application. Many years later, in 1952, a paper was published by Moscona, in which he described the disaggregation of embryonic limb-buds by treating them with trypsin. The cell suspensions could subsequently be cultured. In the same year Dulbecco described a method for preparation of replicate cultures of chick embryonic tissues by digesting the whole embryo with trypsin. The technique has now been applied to a number of embryonic and adult tissues. In particular, it has been used to prepare suspensions of kidney and amnion tissue for virological studies.

Intercellular materials differ in different tissues from mucopolysaccharides to fibrous proteins and include, in addition to carbohydrate and protein elements, inorganic salts as in bone. Thus it might be expected that different tissues would require different treatment in order to obtain disaggregation. In adult tissues this is generally true probably due to the development of large amounts of intercellular materials of a specialized nature. However, embryonic tissues and adult and neoplastic tissues with little matrix can very often be disaggregated by one of three techniques or a combination of these. These are :

1. Physical disruption.
2. Enzymatic digestion.
3. Treatment with chelating agents.

Physical disruption is not often used alone, mainly because it is difficult to obtain uniform suspensions without cell damage. Sometimes, however, it can be used by itself. Thus viable crude suspensions of chick embryonic tissue can be obtained by expressing the embryo through the nozzle of a syringe. Also Lasfargues and Ozzello have obtained viable suspensions of cells from human breast carcinomas by slicing and chopping tissue fragments with a very sharp knife to release the cells from the fibrous stroma. However, physical disruption is usually combined with one of the other methods mentioned below. For instance, before treating tissues with enzymes or chelating agents they are usually cut up into very small fragments while the treatment itself is usually accompanied by fairly vigorous agitation to help break up the tissue.

Enzymatic digestion of the many enzymes that have been employed to disaggregate tissues, trypsin is by far the most commonly used. Others which have been tried successfully are collagenase, elastase, a mucase, trypsin, pancreatin, papain and a preparation from snail livers.

Reports concerning the effectiveness of collagenase are contradictory and this is almost certainly because of the difficulty of obtaining a standard preparation. Lasfargues used it successfully to obtain cultures of normal mammary epithelium from the mouse although it should be noted that complete tissue disaggregation was not obtained in this instance.

In his exhaustive investigation of enzymes capable of disaggregating tissues, Rinaldini found elastase generally ineffective except in the unkeratinized skin of the young (13-day) chick embryo. Although he found *hyaluronidase* almost completely ineffective, he obtained good disruption with a preparation of a mucolytic enzyme from the pancreas.

With regard to *trypsin* and *pancreatin* (a crude extract of pancreatic enzymes) Rinaldini found purified trypsin more effective for separating cells in established cultures but pancreatin more effective for disaggregating tissues from the animal. When pure crystalline trypsin is used for this purpose the cells tend to become stuck together with long mucinous strings. Pancreatin probably contains some mucases which digest this material. Auerbach and Grobstein, on the basis of independent observations, recommended a mixture or trypsin and pancreatin.

Papian, a mixture of *protease* from the papaya plant which is commonly used as a beef tenderiser by virtue of its ability to digest intercellular proteins, has been used effectively as 1/1,000 solution in BSS containing 0.2 per cent cysteine hydrochloride (which is necessary to activate the enzyme preparation).

For the disaggregation of insect tissue Martignoni, Zitcer and Wagner have prepared an enzyme extract from the livers of snails. This extract contains many proteolytic enzymes, including chitinases and viable cell suspensions have been obtained after treatment with it.

Chelating agents

Certain tissues, especially epithelial tissues, seem to require divalent cations, particularly calcium and magnesium, for their integrity. If these ions are removed by substances which bind them the tissue may disrupt very easily. Chelating agents most commonly used are citrate and ethylene-diamine tetra-acetic and (E.D.T.A., Versene or sequestrene). Although Melnick has used Versene to disaggregate fresh kidney tissue these agents are rarcly used for fresh tissues, their main application being the production of cell suspensions from established cultures of epithelial type. A notable exception to this, however, is the technique devised by Anderson for obtaining suspensions of liver cells. He perfuses the liver with citrate and then completes the disruption by forcing the tissue through gauze. This technique has not yet been used with complete success in producing cell-strains but it holds promise for the future.

Trypsin is the only substance used very extensively for disaggregating tissue and some of its applications will be described. It has to be emphasized that not all tissues are attacked by it. Most embryonic organs are easily disaggregated but adult organs with a high content of fibrous tissue are refractory. However, adult tissues with a rather low fibrous tissue content, such as the kidney, are quite easily treated. The original technique described by Moscona is as follows :

Disaggregation of Embryonic limb-buds

Limb-buds from 4-5 day chick embryos are washed repeatedly in balanced salt solution and then in salt solution lacking calcium and magnesium (NaCl–8.00, KCl–0.20g, $NaH_2.PO_4.H_2O$–0.005g., $NaHCO_3$–1.00g., glucose–2.00 g. in 1 liter water). They are transferred to a solution of 3 per cent trypsin (B.D.H.) in this BSS and left for 10-12 minutes at 38°C. The epidermis can then be separated from the underlying tissue and the myogenic from the chondrogenic blastema. The separated parts are cut into smaller pieces and replaced in trypsin solution. By means of 1 per cent KOH solution the pH is raised to 8.4–8.6. The cells become disaggregated after 15-20 minutes, requiring only a little pipetting to complete the process. After washing by taking up in salt solution and centrifuging they are suspended in growth-promoting medium and allowed to settle on glass.

The process described involves fairly rough treatment for the cells and most of the procedures used now are decidedly gentler. Thus, for chick embryonic disaggregation Auerbach and Grobstein recommended using a 3 per cent solution of a mixture of three parts trypsin (Difco 1/250) and one part pancreatin (4 × USP) in Tyrode's solution from which calcium and magnesium salts have been omitted. Ten to twelve minutes swirling at 37.5°C is enough to obtain disaggregation. In particular, when trypsinization procedures are used as a preliminary to cloning cells it is recommended that they should be as gentle as possible. The following procedure is employed successfully for some biochemical studies.

Preparation of Trypsinized Embryonic 'Carcass'

Fourteen to fifteen day chick embryos are decapitated and eviscerated. The skin, wings and feet are removed. The remaining tissue is chopped up quite finely with scissors and washed repeatedly with balanced salt solution to remove erythrocytes. Four or five volumes of 0.5 per cent trypsin (1/250 Difco) are added to the tissue pulp. This is conveniently placed in a roller-tube and left to rotate at 38°C, the pH having been adjusted to about 7.6. After about twenty minutes the tube is removed from the roller drum and the contents sucked up and down once or twice by means of a wide-bore pipette. The contents are allowed to settle for a minute to allow the bigger pieces of debris to separate out. The crude supernatant is transferred to a tube. The tube is then centrifuged and the trypsin discarded. After resuspending in a growth medium the material is inoculated into suitable vessels. The resulting culture contains a great deal of debris. After 48 hours the medium is removed and replaced with 0.5 per cent trypsin. In about ten minutes the cells are again suspended. This time the suspension is filtered through a fine stainless steel gauze screen before centrifuging, resuspending and reinoculating into fresh vessels. The resulting culture consists mainly of muscle cells.

The most general application of this type of technique is in the preparation of monolayers of amnion or kidney epithelium cells for viruses studies. Since very large numbers of cells are required, the technique has been modified mainly by the introduction of mechanic devices to carry out some or all of the process automatically. As in the preceding techniques, disaggregation of the cells is usually achieved by means of trypsin (0.25-0.5 per cent. Difco 1/250) but versene, a strong chelating agent has been shown to be equally effective by Melnick's group and is sometimes preferred. Versene is diamino-ethane-tetra-acetic acid and it is also called E.D.T.A. (ethylene-diamine-tetra-acetic acid). It is used at a concentration of 200 mg. per litre in a buffered solution free of calcium and magnesium ions. The disodium salt is soluble. In the procedure to be described it can be substituted for trypsin but it should be noted that, if this is done, all washing must be performed with calcium and magnesium-free salt solutions.

Fig. 6.3. Trypsinization vessels (a) for intermittent trypsinization, (b) for continuous trypsinization.

The mechanical devices referred to above are all based on the use of an Erlenmeyer flask with indentations in each side to promote turbulence when the contents are agitated by a magnetic stirrer. A further improvement is the flask consists of either a transverse ridge on one side near the neck or a belly blown out on one side. These devices are intended to retain large particulate matter when the cell suspension is decanted off. The fully developed vessel consists of a flask with indentations as before and with an outlet tube coming off underneath in the centre of the floor. A perforated disc at this outlet prevents large particles from failing through into the tube and the centrifugal motion of the stirrer throws most of the particles away from it and towards the periphery. A constant flow of trypsin solution

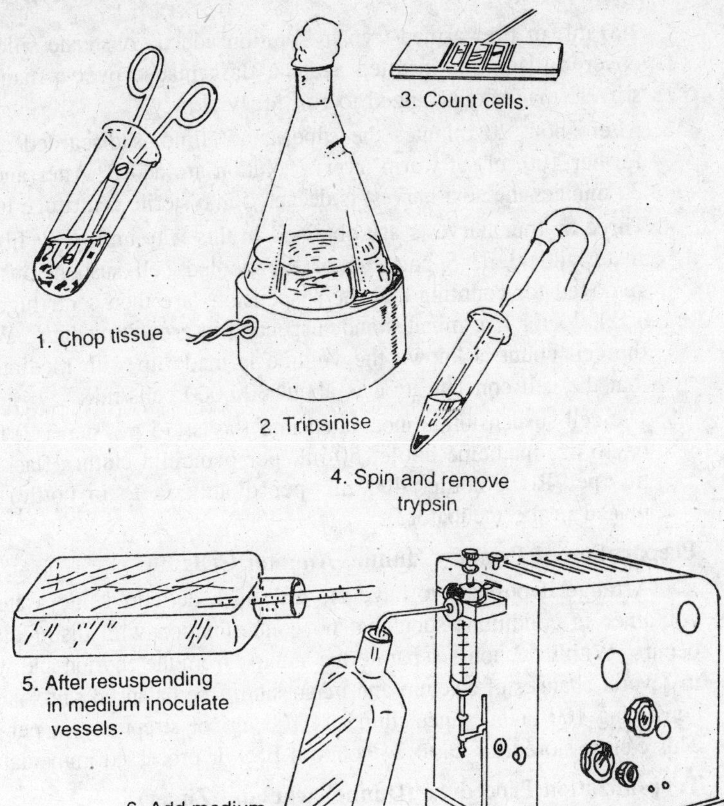

Fig. 6.4. Trypsinization of fresh tissue.

is allowed to enter the flask while a constant flow of cell suspension leaves through the outlet tube.

Trypsinization of Monkey Kidney Tissue

1. The kidneys are removed aseptically either immediately after death or by nephrectomy under anaesthesia before the animal is killed. They are immediately decapsulated and the pelvis is dissected away. The kidney tissue is then chopped up in BSS by means of long-handles scissors until the pieces have a size of 3-5 mm.
2. The BSS is decanted off, replaced with more BSS and this is again decanted off. 5-10 g. of tissue fragments are placed in a 500 ml. Erlenmeyer flask of the type discussed above and about

100 ml. of prewarmed trypsin solution added. A sterile silicone-covered magnet is inserted and the flask placed over a magnetic stirrer (magmix) arranged to run fairly slowly.

3. After about 20 minutes the supernatant fluid is discarded and a further 100 ml. of warm trypsin solution are added. After another 20 minutes the next harvest is decanted into sterile centrifuge tubes.
4. Three or four harvests are obtained in this way until only fibrous material is left. Specimens of the pooled cell suspensions are removed for counting and the suspensions are then centrifuged at 1,000 g. for five minutes and suspended in growth medium. When the cell count is known the volume is made up with medium so that the cell concentration is about 300,000 cells/ml.
5. The cell suspension is inoculated into flasks (15 ml. per T60 flask or 16 oz. medicine bottle, 50 ml. per penicillin culture flask. 80 ml. per Roux bottle, 400 ml. per diphtheria toxin bottle) and placed in the incubator.

Preparation of Primary Human Amnion Cells

Arrange if possible to have the afterbirth delivered into a sterile container in conditions such that no contamination with disinfectants occurs. Within 12 hours separate the amnion from the chorion and wash in several changes of calcium and magnesium-free balanced salt solution containing 100 units of penicillin and 100 μg. of streptomycin per ml. This can be stored for up to 24 hours in BSS or processed immediately.

Trypsinization Procedure (Dunnebacke and Zitcer)

1. Spread the membrane out in a large Petri dish with 50 ml. of 0.25 per cent, trypsin (Difco 1/250) in calcium and magnesium free BSS.
2. After one hour transfer the membrane to a 750 ml. Erlenmeyer flask containing 200 ml. of 0.25 per cent. trypsin solution. Shake manually or mechanically.
3. After 2-4 hours the cells are freed from the membrane, which should be washed with two or three changes of buffer.
4. Combine the cell suspensions from stages 2 and 3 centrifuge at 1,000 r.p.m. for ten minutes. Resuspend in medium and count the cells (a membrane yields 2-6 $\times$ 10^8 cells).
5. Dilute the cells with growth medium (20 per cent, human serum in medium 199) to give 10^6 cells/ml. and inoculate growth vessels.
6. Cultures are confluent after 5-10 days incubation.

The medium requires to be changed every 7 days. Secondary cultures can be prepared by treatment with trypsin or versene but these cells only rarely give rise to permanent strains.

7

PHYSICAL METHODS OF CELL SEPARATION

While cloning or selective culture conditions are the preferred methods for purifying a culture, there are occasions when cells do not grow with a high enough plating efficiency to make cloning possible or when appropriate selection conditions are not available. It may then be necessary to resort to a physical separation technique such as rate or density sedimentation. Physical separation techniques have the advantage that they give a high yield more quickly than cloning although not with the same purity.

The more successful separation techniques depend on differences in: (1) cell size; (2) cell density (specific gravity); (3) Cell surface charge;

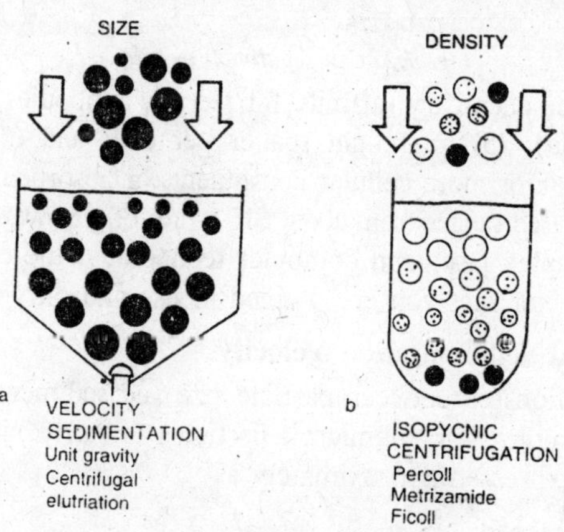

Fig. 7.1. Cell separation techniques.

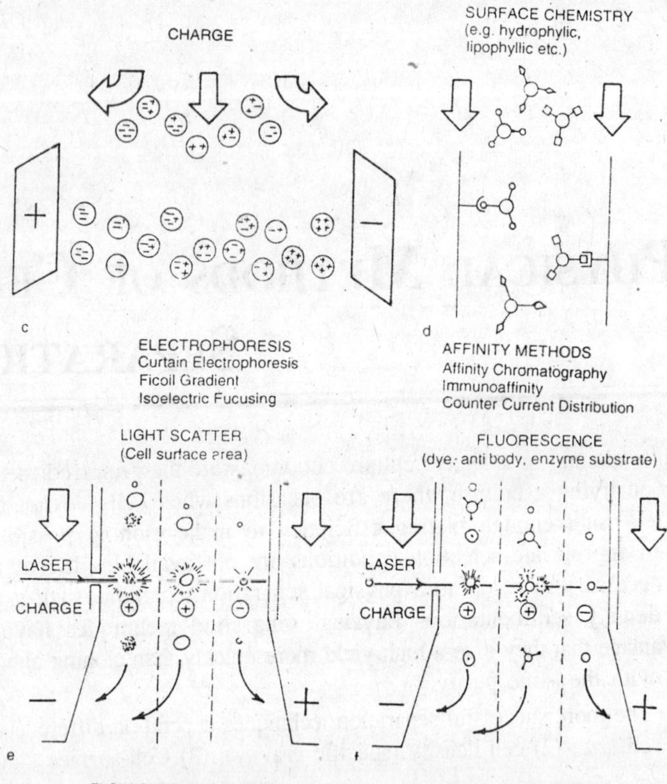

Fig. 7.2. Cell separation techniques.

(4) cell surface chemistry (affinity for lectins, antibodies, or chromatographic media); (5) total light scatter per cell; and (6) fluorescence emission of one or more cellular constituents or absorbed antibody. The apparatus required ranges from about \$10-worth of glassware to \$ 200,000-worth of complex laser and computer technology; the choice depends on the parameter that you are obliged to use and on your budget.

Cell Size and Sedimentation Velocity

The relationship between particle size and sedimentation rate at 1 g though complex for submicron particles is fairly simple for cells and can be expressed approximately as

$$v = \frac{r^2}{4} \qquad ...(1)$$

where v=sedimentation rate in mm/hr and r=radius of the cell in μm.

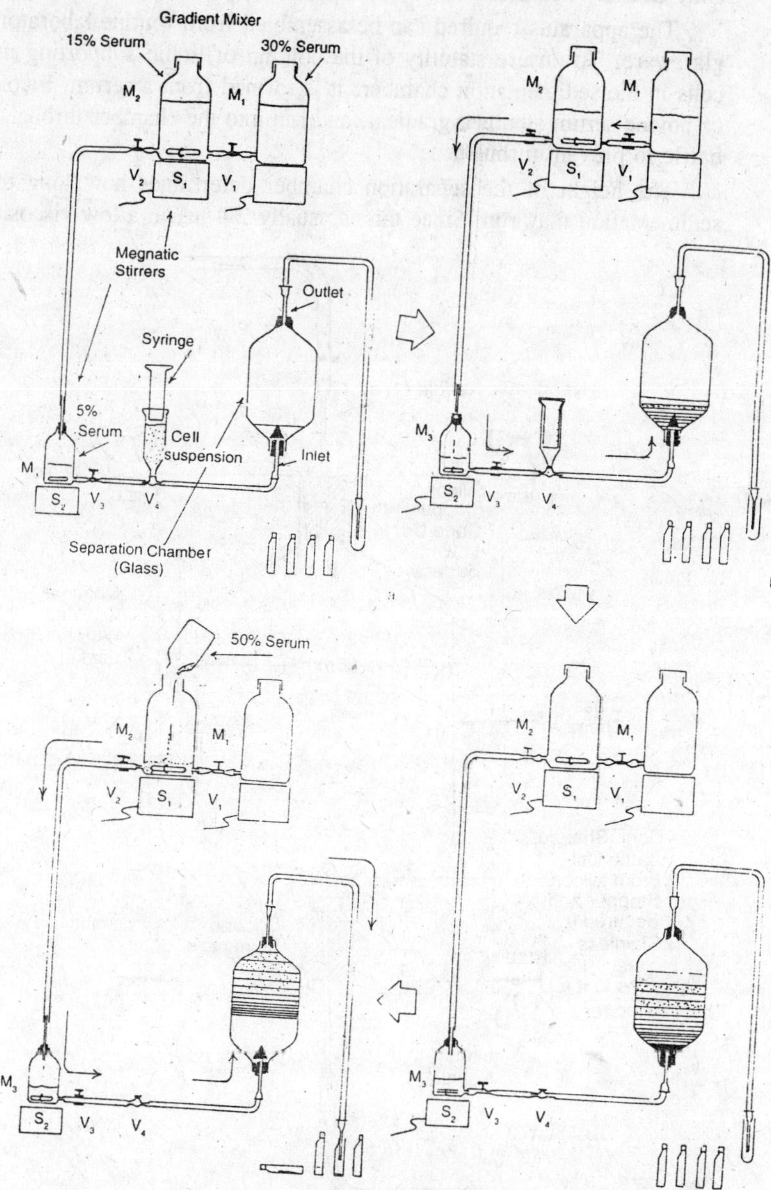

Fig. 7.3. Cell separation velocity sedimentation at unit gravity. Apparatus and position of valves at different stages of the procedure. (a) At start, loading cells. (b) Running in gradient. (c) After cell sedimentation. (d) Harvesting. V–valve; M—mixer vessel; S—stirrer.

Unit Gravity Sedimentation

The apparatus required can be assembled from routine laboratory glassware. To ensure stability of the column of liquid supporting the cells in the sedimentation chamber, it is formed from a serum, Ficoll, or bovine serum albumen gradient, and run into the chamber through a baffle to prevent turbulence.

The height of the separation chamber determines how long the sedimentation may run. Since this is usually 2-4 hr, in a low-viscosity

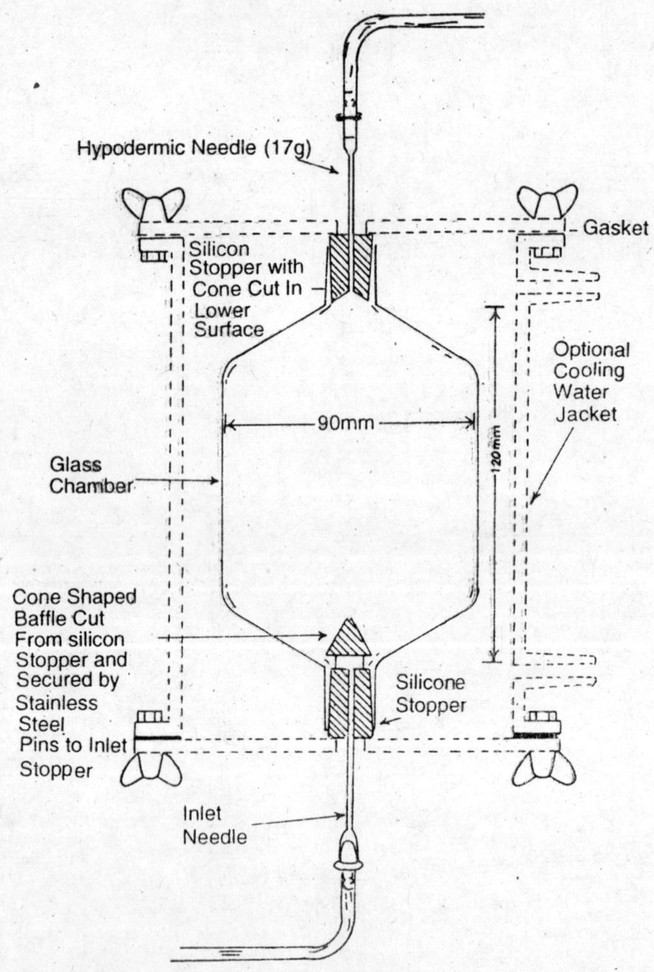

Fig. 7.4. Separation chamber for unit gravity sedimentation. The dotted outline is a cooling jacket for carrying out sedimentation at 4°C.

medium, 10 cm is approximately correct. A longer sedimentation time may give better resolution but may cause deterioration of the cells.

The width of the chamber controls the number of cells that may be loaded onto the gradient as the cell layer should be kept thin (~ 5 mm) and the cell concentration low (~ 106 ml).

The height and width (i.e., volume) also affect the filling rate. The chamber must not be filled too rapidly or turbulence will result, and it cannot be filled too slowly or the cells will sediment faster than the liquid level rises. The dimensions are optimal for separating about 2×10^7 cells of 15-18 μm diameter or up to 10^8 cells of 10-12 μm diameter.

The procedure for separating a typical cell suspension of average cell diameter 15μm is as follows.

Outline

Float cells on top of a gradient of serum in medium, allow cells to sediment through the gradient for about 3 hr, and run off gradient into culture vessels.

Materials

300 ml Eagle's MEMS (suspension salts) + 30% serum.

300 ml MEMS + 15% serum

20 ml MEMS + 5% serum

10 ml. 0.25% trypsin-citrate

30 ml PbS

20 ml MEMS + 3% serum

hemocytometer or cell counter flotation medium (1M sucrose or 20% Ficoll)

25-cm^2 flsks

growth medium

Protocol

1. Prepare apparatus. Incorporate Luer connections to allow for disassembly for sterilization. Package and autoclave.
2. Assemble and check that valves V_1, V_2, V_3 and V_4 are closed.
3. Add 300 ml 30% serum is medium to mixer vessel M_1 and 300 ml 15% serum in medium to mixer vessel M_2.
4. Check that stirrer S_1 is functioning.
5. Add 20 ml 5% serum in medium to mixer M_3 and check that stirrer S_2 is functioning.

6. Open V_4 to connect syringe to separation chamber and insert 20 ml PBS into separation chamber.
7. Opern V_4 to M_3 line, open V_3, and dray a little 5% serum into the syringe (just enough to fill line). Close V_3 and V_4.
8. Prepare cell suspension, e.g., by trypsinizing primary culture for 15 min in 0.25% trypsincitrate. Disperse cells carefully in 3% serum in medium and check that a single cell suspension is formed.
9. Take up 20 ml at 106 /ml (maximum) into syringe and connect to V4 inlet.
10. With syringe held vertically, open V_4 to M_3 line, open V_3, and draw a littler 5% serum into syringe to clear any bubbles from M_3 line and V_4.
11. Turn valve V_4 to separation chamber line and draw a little PBS into syringe to clear any bubbles from this line.
12. Insert cell suspension slowly into chamber; avoid mixing cell suspension with the overlaying PBS layer. Take care to stop while a little fluid is left in the syringe to avoid injecting any air bubbles back into the line. If difficulty is encountered injecting cells smoothly, without turbulence, remove piston from syringe and allow cells to run in under gravity alone by raising V_4.
13. Start stirrers S_1 and S_2.
14. Open V_4 to connect M_3 line to separation chamber.
15. Open V_1 and adjust flow rate by opening V_2 to give 15 ml./min (~ 5 drops/s) at M_3. Cell suspension will now float up into separation chamber on gradient of serum. Check for turbulence at baffle as suspension and gradient run it. If there is any, reduce flow rate at M_3 by closing V_2.
16. When gradient mixers M_1 and M_2 are empty, but before M_3 empties, close V_3 and V_2.
17. It should be possible to see the cell layer in the sedimentation chamber and to follow the cells as they sediments. As they do, the cell band will become wider and more diffuse. Check for signs of "streaming" in the early stages of sedimentation (tails of cells which sediment ahead of the main band). This occurs when the cell concentration is too high or the step between the cells and the gradient is too steep.
18. After about 20 min, close V_1 and add 90 ml 50% serum to M_2. Open V_3 and adjust flow rate at M_3 to 15 ml/nin. Stop when M_2 is empty but before M_3 empties by closing V_3 and V_2.

19. Add 500 ml. flotation medium (1 M sucrose or 20% Ficoll) to M_1 and M_2, open V_1 and V_2, and let some of the flotation medium run into M_3. Close V_2.
20. When sedimentation is complete, i.e., cell band midway down separation chamber, open V_3, and adjust V_2 to give a flow rate of 15 ml/ min in M_3.
21. Collect eluate from top of chamber via elution line and run into graduated culture vessels, e.g., 25-cm^2 flasks, 10 ml per flask. Mix the contents of each flask and take sample for cell counting.
22. Seal and incubate flasks for 24 hr, replace medium with fresh medium at standard serum concentration and volume.

Variations

Gradient medium. If serum and regular culture medium are used, then it is possible to culture cells directly from the eluate. Fetal bovine serum causes less reaggregation than calf or horse serum. If serum is found to be unsuitable, gradients can be formed from bovine serum albumen or Ficoll (Pharmacia).

Aggregation. Aggregation can be reduced by enclosing the separation chamber in a cooling jacket and running the whole process at 4°C. Mixers M_1, M_2 and M_3 must all be kept cold also. Water-driven

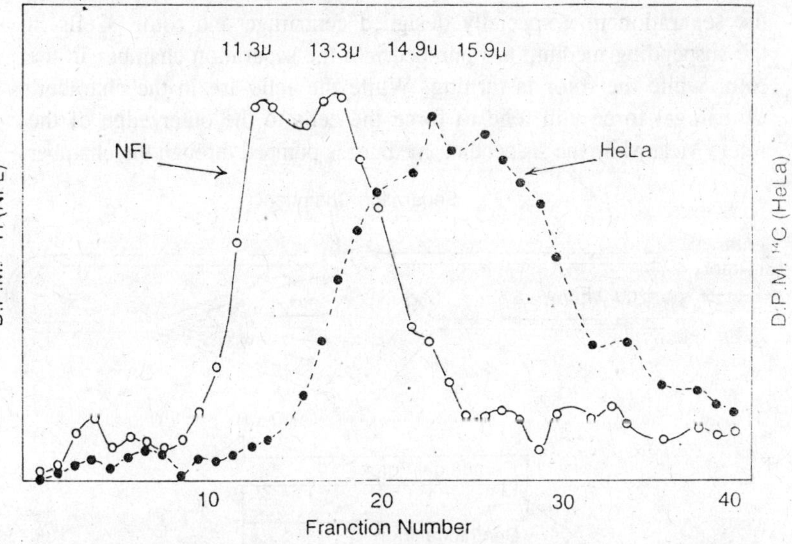

Fig. 7.5. Elution profiles of artificial mixture of HeLa and NFL (normal lumen lung fibroblasts) after sedimentation at 1 g for 3 hr.

magnetic stirrers may be used at S_1 and S_2 to minimize overheating of gradient.

Pump. Gravity is used in this example as the cheapest and simplest method for generating the flow of gradient medium, but if desired, a peristatlic pump may be inserted at V_3.

Sedimentation of cells at unit gravity is a simple low-technology method of separating cells. It works well for many cell types, e.g., brain hemopoietic cells and HeLA/fibroblast mixtures and can be performed in regular physiological media. The cells must be singly suspended, however and there is a practical limit of about 10^8 cells that may be separated.

Pretlow and others have used specially formed gradients of Ficoll to separate cells by sedimentation velocity at higher g forces on a zonal rotor. The gradients are shallow and of relatively low density to minimize the effect of cell density on sedimentation rate. Cells of may different types have been separated by this method, and it appears to have a wide application.

Centrifugal Elutriation

The centrifugal elutriator is a device for increasing the sedimentation rate and improving the yield and resolution by performing the separation in a specially designed centrifuge and rotor. Cells in the suspending medium are pumped into the separation chamber in the rotor while the rotor is turning. While the cells are in the chamber, centrifugal force will tend to force the cells to the outer edge of the rotor. Meanwhile the suspending medium is pumped through the chamber

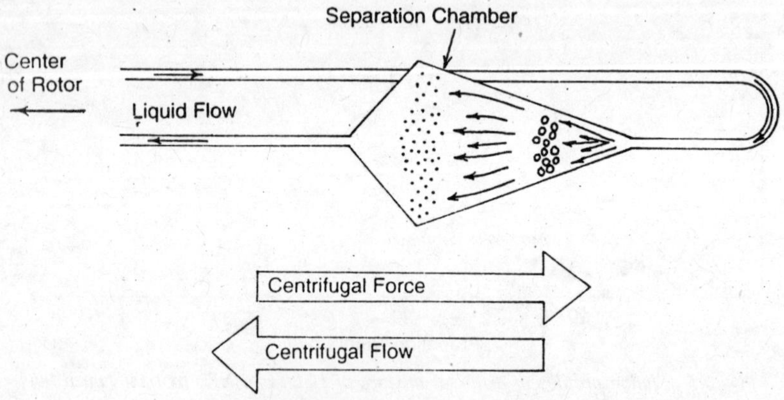

Fig. 7.6. Separation chamber of elutriator rotor.

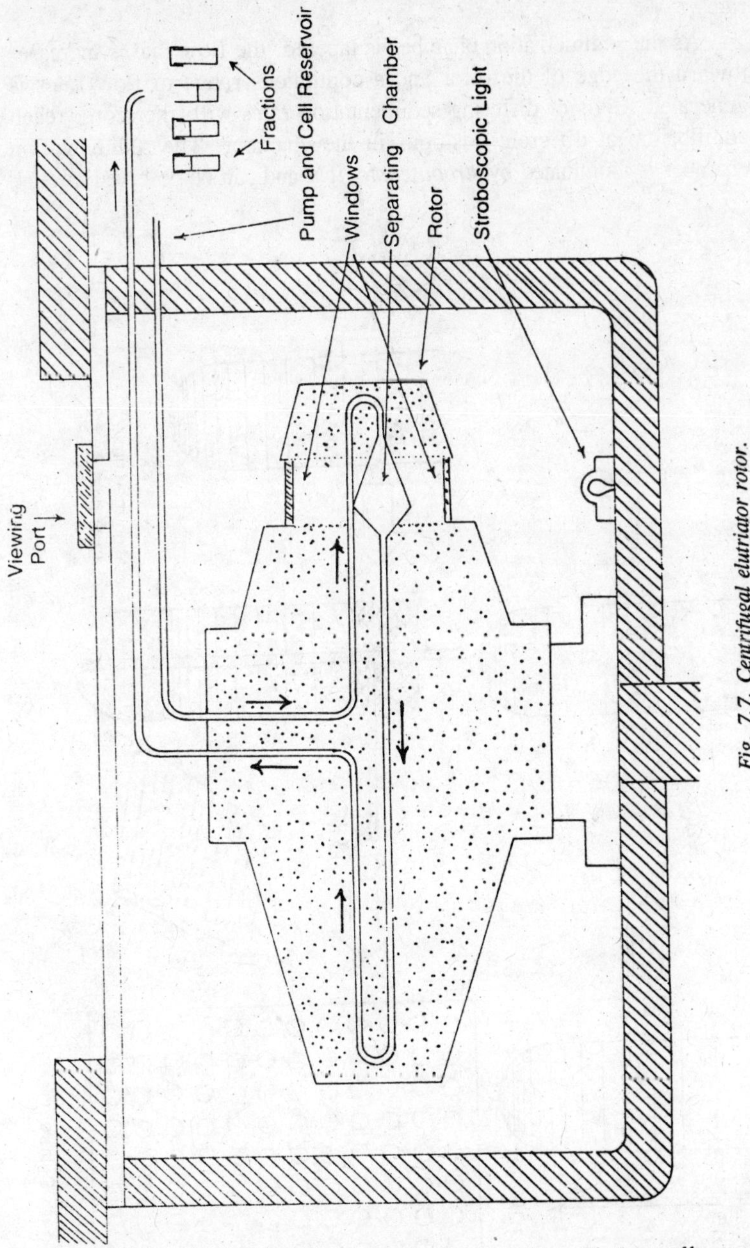

Fig. 7.7. Centrifugal elutriator rotor.

such that the centripetal flow rate of the cells. If the cells were uniform, they would remain stationary, but since they vary in size, density and cell surface configuration, they tend to sediment at different rates.

As the sedimentation chamber is tapered, the flow rate is increases toward the edge of the rotor and a continuous range of flow rates is generated. Cells of differing sedimentation rates will, therefore, reach equilibrium at different positions in the chamber. The sedimentation chamber is illuminated by *stroboscopic light* and can be observed through

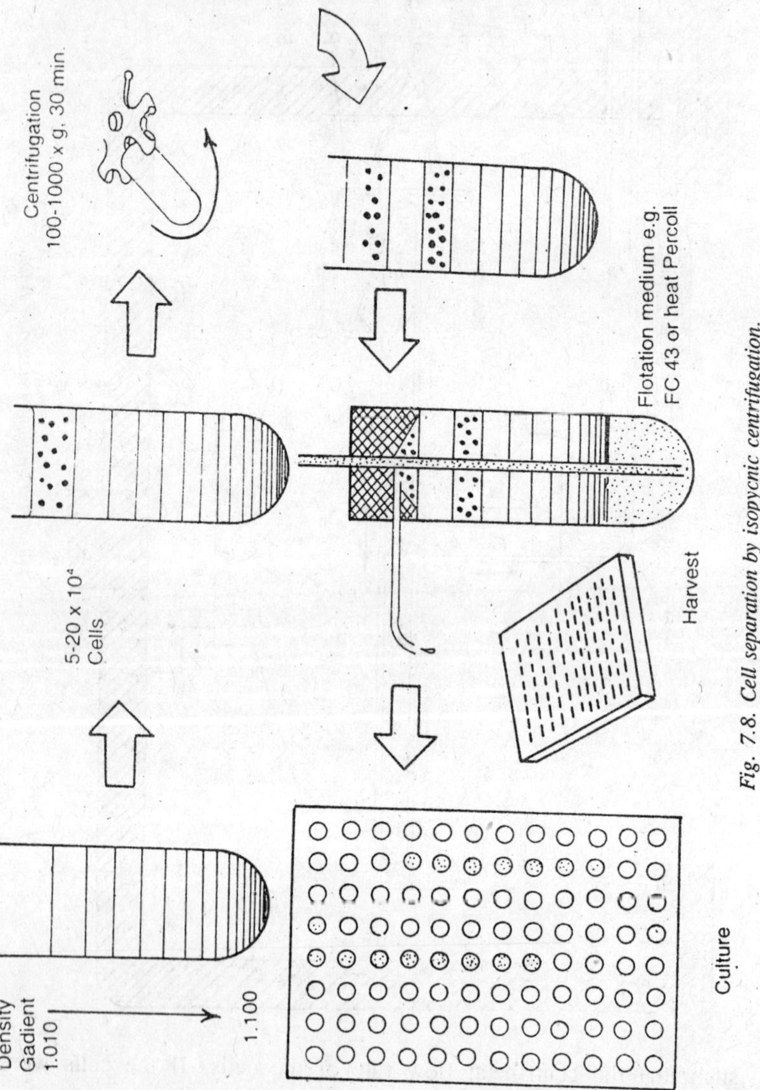

Fig. 7.8. Cell separation by isopycnic centrifugation.

a viewing port. When the cells are seen to reach equilibrium, the flow rate is increased and the cells are pumped out into receiving vessels. The separation can be performed in complete medium and the cells cultured directly afterward.

Equilibrium is reached in a few minutes and the whole run may take 30 min. On each run 10^8 cells may be separated and the run may be repeated as often as necessary. The apparatus is, however, fairly expensive and a considerable amount of experience is required before effective separations may be made. A number of cell types have been separated by this method as have cells of different phase of the cells cycle.

Cell Density and Isophycnic Sedimentation

Separation of cells by density can be performed at low or high *g* using conventional equipment. The cells sediment in a density gradient to an equilibrium position equivalent to their own density (*isopycnic sedimentation*). Physiological media must be used and the osmotic strength carefully monitored. The density medium should be nontoxic, nonviscous at high densities (1.10 g/ ml) and exert little osmotic pressure in solution. Serum albumen, dextran, Ficoll metrizamide (Nygaard) and Percoll (Pharmacia) have all been used successfully; Percoll (colloidal silica) is one of the more effective media currently available.

Outline

Form gradient; (1) by layering different densities of Percoll; (2) by high-speed spin; or (3) with special gradient former. Centrifuge cells through Percoll gradient (or allow to sediment at unit gravity), collect fractions, and culture directly.

Materials

Culture medium (sterile)
medium + 20% Percoll (sterile)
25-ml centrifuge tubes (sterile)
PBSA (sterile)
0.25% trypsin (sterile)
syringe or gradient harvester (sterile)
24-well plates or microtitration plates (sterile)
refractometer or density meter
hemocytometer
cell counter

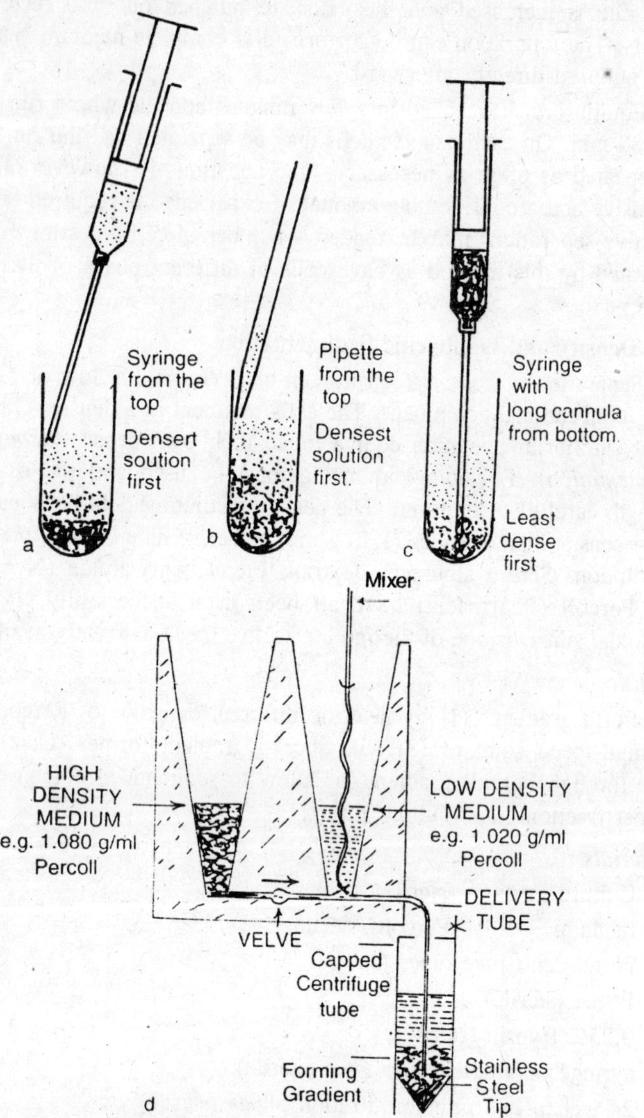

Fig. 7.9. Layering density gradients (a) by syringe from the top, (b) by pipette from the top, (c) by syringe from the bottom, (d) by gradient mixing device.

Protocol

1. *Prepare gradient*: (1) Prepare two media, one regular culture medium and one with 20% Percoll; (1) adjust the density of the

Percoll solution to 1.10 g/ml and its osmotic strength to 290 mOsm kg; (3) mix the two media in varying proportions to give the desired density range (e.g., 1.020-1.100 g/ml) in ten or 20 steps; and (4) layering one step over another, build up a stepwise density gradient in a 25-ml centrifuge tube. Gradients may be used immediately or left overnight.

Alternatively, place medium containing Percoll of density 1.085 g/ml in a tube and centrifuge at 20,000 g for 1 hr. This generates a sigmoid gradient, the shape of which is determined by the starting concentration of Percoll, the duration and centrifugal force of the centrifugation, the shape of the tube, and the type of rotor. A continuous linear gradient may be produced by mixing, for example, 1.020 g/ml with 1.08 g/ml Percoll in a gradient-forming device.

2. Trypsinize cells and resuspend in medium plus serum. Check that they are singly suspended.
3. Layer up to 2 × 10^7 cells in 2-ml medium on top of the gradient.
4. The tube may be allowed to stand on the bench for 4 hr or centrifuged for 20 min at between 100 and 1,000 g.
5. Collect fractions using a syringe, or a gradient harvester (MSE/Fisons). Fractions of 1 ml may be collected into a 24-well plate or 0.1 ml into microtitration plates. Samples should be taken at

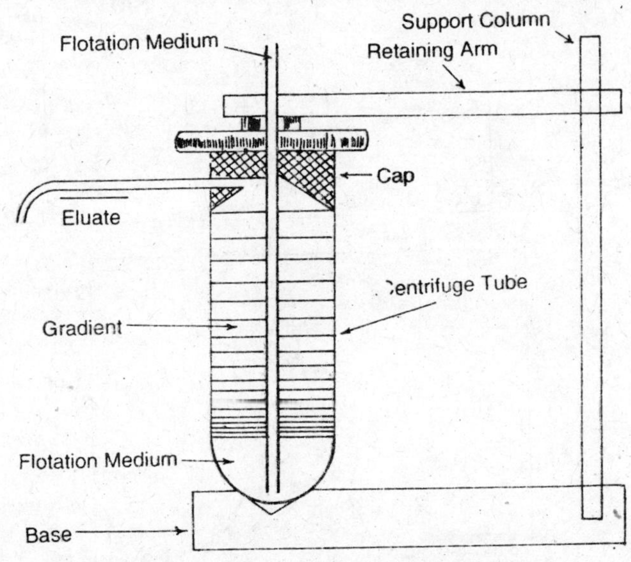

Fig. 7.10. Gradient harvester (Fisons/M.S.E.).

intervals for cell counting and determination of the density (r) of the gradient medium. Density may be measured on a *refractometer* (Hilger) or density meter (Paar).

6. Add equal volume of medium to each well and mix (to ensure cells settle to bottom of well). Change the medium to remove the Percoll after 24-28 hr incubation.

Variations

Cell may be incorporated into the gradient during formation by centrifugation. Only one spin is required although spinning the cells at such a high *g* force may damage them.

Other media

Ficoll is one of the most popular media as it, like Percoll, can be autoclaved. It is a littler more viscous at high densities and may cause agglutination of some cells. Metrizamide (Nygaard), a nonionic

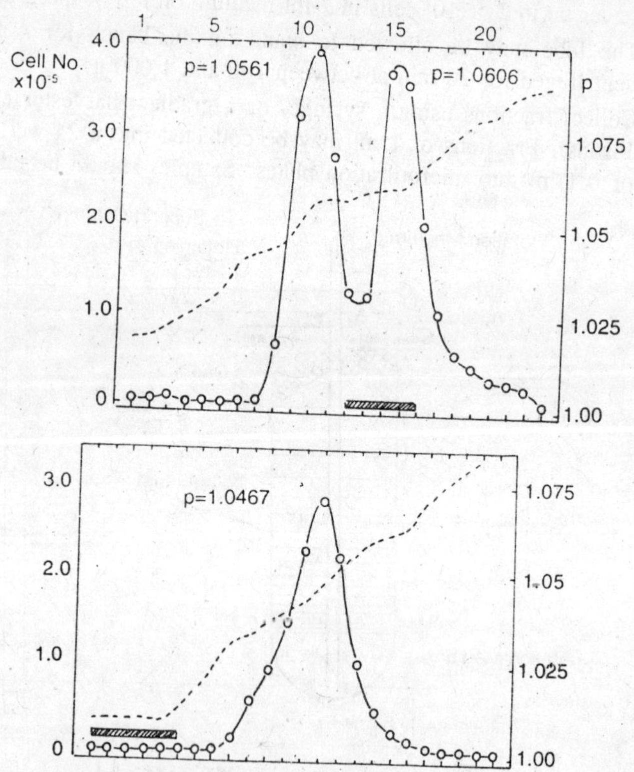

Fig. 7.11. Incorporation of metrizamide (Nygaard) into cells during isopycnic centrifugation.

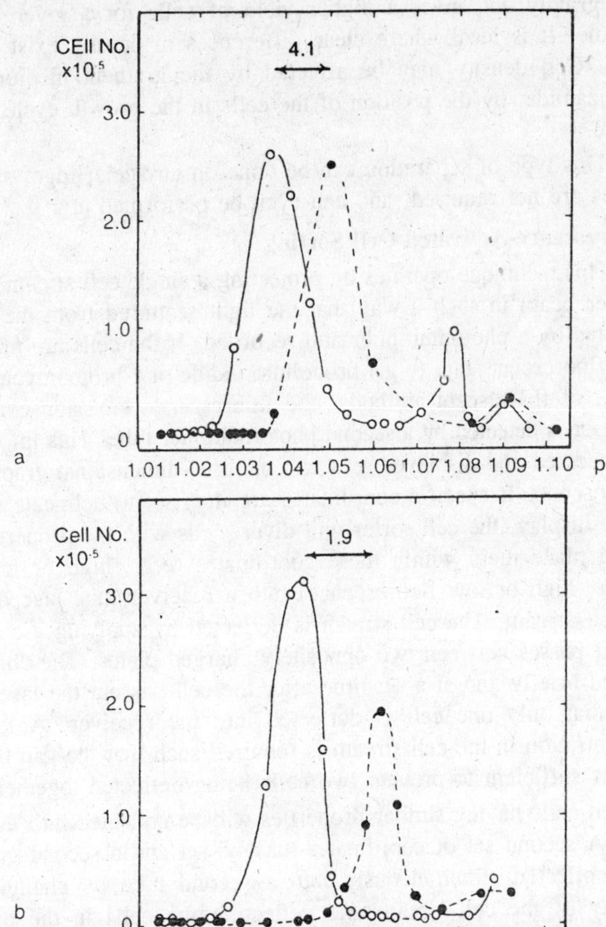

Fig. 7.12. Sedimentation profiles of HeLa and MRC-5 cells, centrifuged to equilibrium in gradients of metrizamide in culture medium, a, b.

derivative of metrizoate, which is a radio-opaque-iodinated substance used in radiography (Isopaque, Hypaque, Renografin) and in lymphocyte purification (e.g., Lymphoprep) is less viscous at high densities but may be incorporated into some cells as is Isopaque. Where such media are used, cells should always be layered on top of the gradient and not mixed in during formation.

Marker beads

Pharmacia manufacture coloured marker beads of standard densities which may be used to determine the density of regions of the gradient.

Isopycnic sedimentation is quicker than velocity sedimentation at unit gravity and gives a higher yield of cells for a given gradient volume. It is ideal where clear differences in density exist between cells. Cell density may be affected by the gradient medium (e.g., metrizamide, by the position of the cells in the growth cycle, and by serum.

This type of separation can be done on any centrifuge, as high *g* forces are not required, and can even be performed at 1 g.

Fluorescence-Activated Cell Sorting

This technique operates by projecting a single cell stream through a laser beam in such a way that the light scattered from the cells is detected by a photomultiplier and recorded. If the cells are pretreated with fluorescent stain (e.g., propidium iodide or Chromomycin A_3 for DNA) of fluorescent antibody, the fluorescence emission excited by the laser is detected by a second photomultiplier tube. This information is processed and displayed as a two or three-dimensional graph on an oscilloscope. If specific coordinates are then set to delineate sections of the display, the cell sorter will divert cells with the properties that would place them within these coordinates (e.g., high or low light scatter, high or low fluorescence) into a receiver tube placed below the cell stream. The cell stream is deflected by applying a charge to it as it passes between two oppositely charged plates. The charges is applied briefly and at a set time after the cell has cut the laser beam such that only one cell is deflected into the receiver. A low cell concentration in the cell stream is required such that the gap between cells is sufficient to prevent two cells being deflected together.

All cells having similar properties will be collected into the same tube. A second set of coordinates may be set and a second group of cells collected simultaneously into a second tube by changing the polarity of the cell stream and deflecting the cells in the opposite direction. All remaining cells will be collected in a central reservoir.

This method may be used to separate cells with any differences that may be detected by light scatter (e.g., cell size) or fluorescence (e.g., DNA, RNA, protein, enzyme activity, specific antigens.) It is an extremely powerful tool but limited by cell yield (about 107 cells is a reasonable maximum) and the very high cost of the instrument (approximately $100,000). It also requires a full-time, skilled operator.

Other Techniques

The many other techniques which have been used successfully to separate cells are too numerous to describe in detail

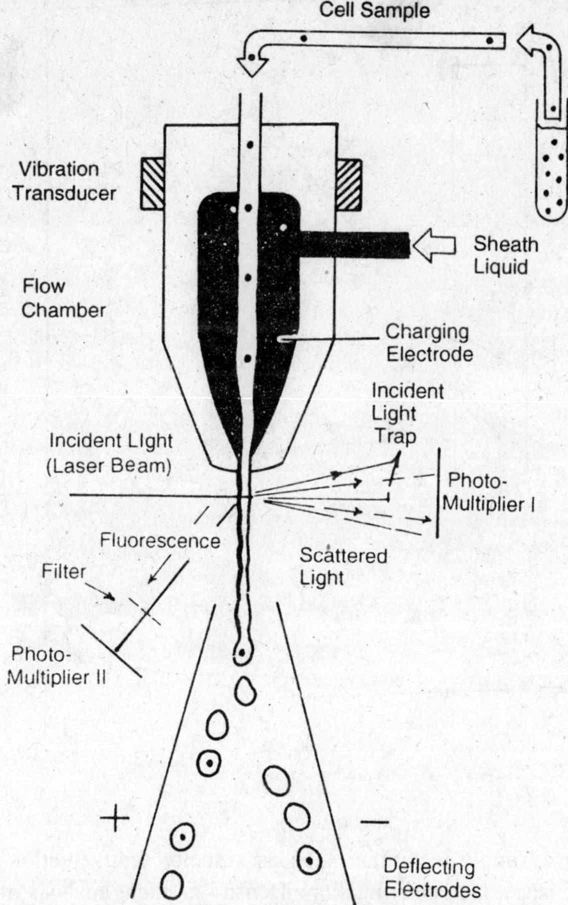

Fig. 7.13. Principle of operation of flow cytophotometer.

Electrophoresis either in Ficoll gradient or by curtain electrophoresis; the second technique is probably more effective and has been used to separate kidney tubular epithelium.

Affinity chromatography on antibody or plant lectins bound to nylon fiber or Sephadex (Pharmacia). These techniques appear to be useful for fresh blood cells but less so for cultured cells.

Counter current distribution has been used to purify murine ascites tumor cells, but the viability may be too low for subsequent culture.

As so many techniques exist, it is difficult for the novice to know where to start. Like so many other areas of investigation, it is best to

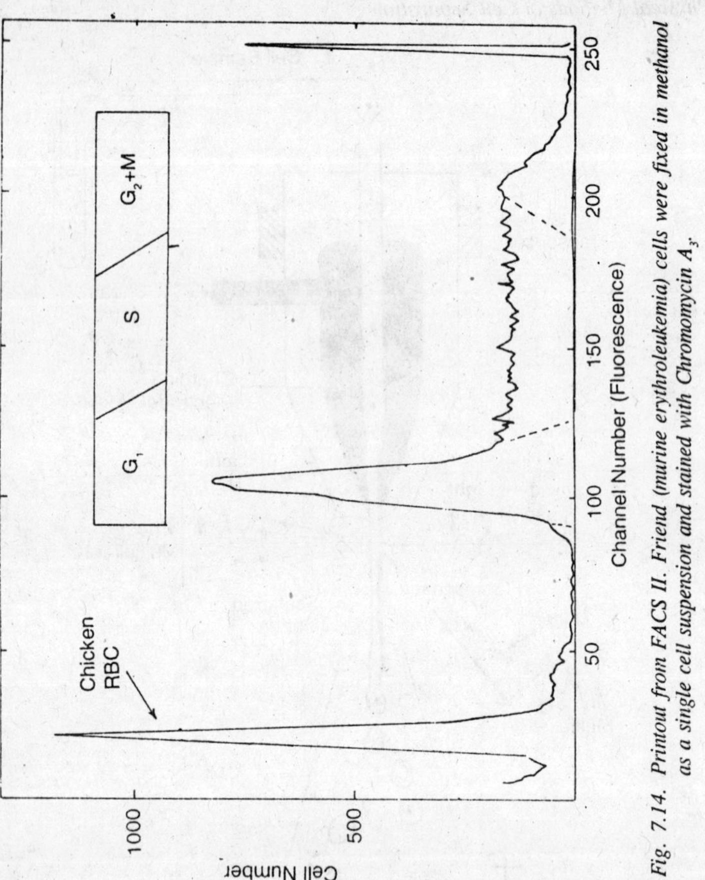

Fig. 7.14. Printout from FACS II. Friend (murine erythroleukemia) cells were fixed in methanol as a single cell suspension and stained with Chromomycin A_3.

start with a simple technique such as velocity sedimentation at unit gravity or isopycnic centrifugation. Density gradient analysis may also be used in conjunction with velocity sedimentation in a two-stage fractionation. If there are still problems of resolution or yield, then it may be necessary to employ high-technology methods such as centrifugal elutriation, curtain electrophoresis, or fluorescence-activated cell sorting.

8

Aseptic Technique

In spite of the introduction of antibiotics, contamination by micro-organisms remains a major problem in tissue culture. Bacteria, mycoplasma, yeasts, and fungal spores may be introduced via the operator, atmosphere, work surfaces, solutions, and many other sources. Contaminations can be minor and confined to one or two cultures, can spread between several and infect a whole experiment, or can be widespread and wipe out your, or even the whole laboratory's, entire stock. Catastrophes can be minimized if: (1) Cultures are checked on the microscope, preferably by phase contrast, every time that they are handled; (2) they are kept antibiotics-free for at least part of the time to reveal cryptic contaminators; (3) reagents are checked for sterility before use; (4) bottles of media etc., are not shared with other people or used for different cell lines; and (5) the standard of sterile technique is kept at all times.

Mycoplasmal infection, invisible under regular microscopy, presents one of the major threats. Undetected, it can spread to other cultures around the laboratory. It is, therefore, essential to back up visual checks with a mycoplasma test, particularly if cell growth appears abnormal.

Objectives of Aseptic Technique

Correct *aseptic technique* should provide a barrier between microorganisms in the environment outside the culture and the pure uncontaminated culture within its flask or dish. Hence, all materials which will come into direct contact with the culture must be sterile and manipulations designed such that there is not direct link between the culture and its nonsterile surroundings.

It is recognized that the sterility barrier cannot be absolute without working under conditions which would severely hamper most routine manipulations. Since testing the need for individual precautions would be an extensive and lengthy controlled trial, procedures are adopted largely on the basis of common sense and experience. Aseptic technique is a combination of procedures designed to reduce the probability of infection, and the correlation between the omission of a step and subsequent contamination is not always absolute. The operator may abandon several precautions before the probability rises sufficiently that a contamination occurs. By then, the cause becomes multifactorial and consequently no simple solution is obvious. If once, established, all precautions are maintained consistently, breakdown will be rather and more easily detected.

Although laboratory conditions have improved in some respects (air conditioning and filtration, laminar flow facilities, etc.) the modern laboratory is often more crowded and accommodation may have to shared. However, with reasonable precautions, maintenance of sterility is not difficult.

Quiet Area

In the absence of laminar flow, a separate sterile room should be used if possible. If not, pick a quiet corner of the laboratory with little or no traffic and no other activity. With laminar flow, an area should be selected which is free from drafts and traffic should still be kept to a minimum. Animals and microbiological culture should be excluded from the tissue culture area. It should be kept clean and dust free and should not contain equipment other than that connected with tissue culture.

Work Surface

One of the more frequent examples of bad technique is the failure to keep the work surface clean and tidy. The following rules should be observed:

1. Start with a completely clear surface.
2. Swab down liberally with 70% alcohol.
3. Bring onto it only those items you require for a particular procedure and swab bottles, cans, etc., with 70% alcohol beforehand.
4. Arrange your apparatus (a) to have easy access to all of it without having to reach over one item to get at another and (b) to leave a wide, clear space in the center of the bench (not just the front edge) to work on. If you have too much equipment too close to

you, you will inevitably brush the tip of a sterile pipette against a nonsterile surface.

5. Work within your range of vision, e.g. insert a pipette in a bulb with the tip of the pipette pointing away from you so that it is in your line of sight continuously an not hidden by your arm.
6. Mop up any spillage immediately and swab with 70% alcohol.
7. Remove everything when you have finished and swab down again.

Personal Hygiene

There has been much discussion about whether hand-washing encourages or reduces the bacterial count on the skin. Regardless of this debate, washing will moisten the hands and remove dry skin likely to blow onto your culture and reduce loosely adherent microorganisms, which are the greatest risk to your cultures. Surgical gloves may be worn and swabbed frequently, but it may be preferable to work without (where no hazard is involved) and retain the extra sensitivity that this allows.

Caps, gowns, and face masks are often worn but are not always strictly necessary, particularly when working with laminar flow. However, if you have long hair, tie it back. When working on the open bench, do not talk while working aseptically; and if you have a cold, wear a face mask, or, better still, do not do any tissue culture during the height of the infection. Talking is permissible when working in vertical laminar flow with a barrier between you and the culture but should still be kept to a minimum.

Pipetting

Standard glass or disposable plastic pipettes are still the easiest form of manipulating liquids. Syringes are often used, but regular needles are too short to reach into moist bottles. Syringing may produce high shearing forces when dispensing cells and increase the risk of self-inoculation.

Pipettes of a convenient size range should be selected—1 ml, 2 ml, 5 ml, 10 ml, and 25 ml cover most requirements. If you only require a few of each, make up mixed cans and save space. Mouth pipetting, even with plugged pipettes or a filter tube/mouthpiece should be avoided, as it has been shown to be a contributory factor in mycoplasmal infection and may introduce an element of hazard to the operator, e.g., with virus-infected cell lines and human biopsy or autopsy specimens and biohazards. Inexpensive bulbs and pipetting devices are available; try a selection of these to find one that suits you. They

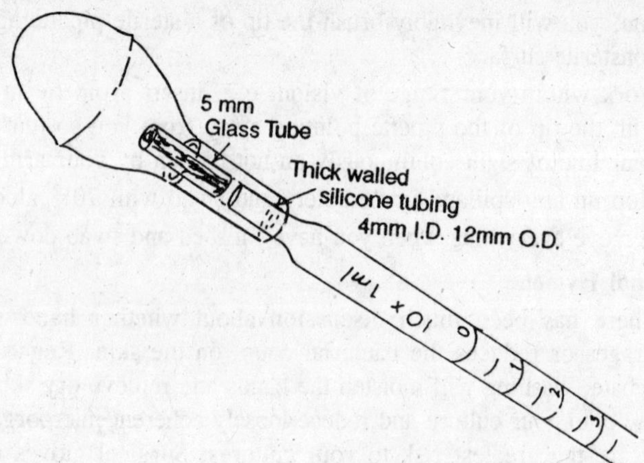

Fig. 8.1. Filter tube. Interposed between bulb and pipettes, avoids the necessity to plug pipettes. Filter tube must be renewed between cell lines or if wetted.

should accept securely all the sizes of pipette that you use without forcing them and without the pipette falling out. The regulation of flow should be easy and rapid, but at the same time, capable of fine adjustment. You should be able to draw liquid up and down repeatedly (e.g., to disperse cells) and there should be no fear of carry-over. The device should fit comfortably in your hand and should be easy to operate with one hand.

The Marburg-type pipette is particularly useful for small volumes (1 ml and less) though there can be some difficulty in reaching down into larger vessels with most of them. They are best used in conjunction with a shallow vial or bottle and are particularly useful when dealing with microtitration assays and other multi well dishes but should not be used for serial propagation. Multipoint pipettors are available for microtitration dishes.

It is necessary to insert a cotton plug in the top of a glass pipette before sterilization to maintain sterility in the pipette during use. If this becomes wet in use, discard the pipette into the washing up. Plugging pipettes for sterile use is a very tedious job, as in the removal of plugs before washing. Automatic pipette pluggers are available, and although expensive, speed up the process and reduce the tedium. Alternatively, a sterile filter tube may be attached to the bulb, eliminating the need to plug pipettes. It is important that the filter tube is changed between handling of different cell lines to avoid the risk of cross contamination.

Sterile Handling

Swab bottles, particularly from the cold room, before using for the first time each day.

Capping

Deep screw caps should be used in preference to stoppers although care must be taken when washing caps to ensure that all detergent is rinsed from behind rubber liners. Wadless polypropylene caps should be used if possible. The screw cap should be covered with aluminum foil to protect the neck of the bottle from sedimentary dust.

Flaming

When working on the open bench, the necks of bottles and screw caps should be flamed before and after opening a bottle and before and after closing. Pipettes should be flamed before use. Work close to the flame where there is an up-current due to convection, and do not leave bottles open. Screw caps should be placed open side down on a clean surface and flamed before replacing on the bottle.

Pouring

Whenever possible, do not pour from one sterile container into another unless the bottle you are pouring from is to be used once only and, preferably, is to deliver all its contents in one single delivery. The major risk in pouring lies in the generation of a bridge of liquid between the outside of the bottle and the inside which may permit infection to enter the bottle.

Laminar Flow

The major advantage of working in *laminar flow* is that the working environment is protected from dust and contamination by a constant stable flow of filtered air passing over the work surface. There are two main types: (1) horizontal, where the air flow blows from the side facing you parallel to the work surface, and is not recirculated; and (2) vertical, where the air blows down from the top of the cabinet on the work surface and is drawn through the work surface and either recirculated or vented. In recirculating hoods, 20% is vented and made up by drawing in air at the front of the work surface. This is designed to minimize overspill from the work area of the cabinet. Horizontal flow hoods give the most stable airflow and best sterile protection to the culture and reagents; vertical flow gives more protection to the operator. If potentially hazardous material (radioisotopes, mutagens, human or primate derived cultures, virally infected cultures etc.) is being handled, a Class II vertical flow biohazard hood should be used.

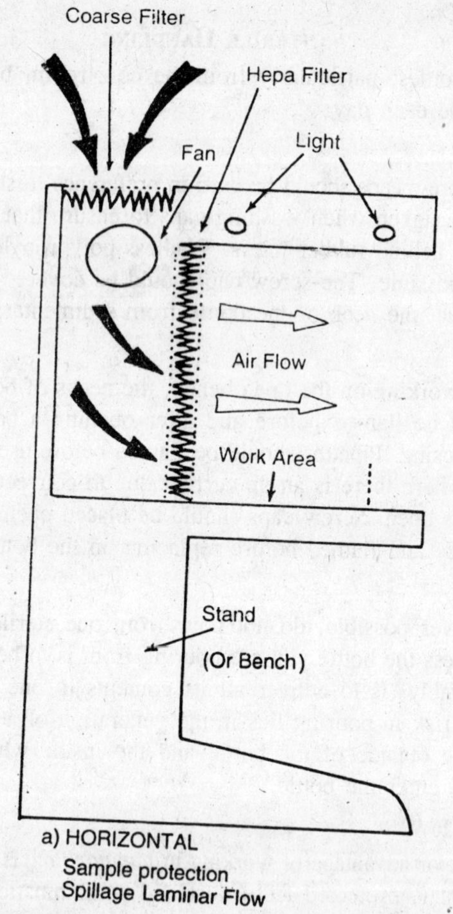

Fig. 8.2. Horizontal laminar flow hoods. Filled arrows, nonsterile air; open arrows, sterile air.

If known human pathogens are handled, a class III pathogen cabinet with a pathogen trap on the vent is obligatory.

Laminar flow hoods depend, for their efficiency, on a minimum pressure drop across the filter. When filter resistance builds up, the pressure drop increases and the flow rate or air in the cabinet falls. Below 0.4 m/s (80 ft/min), the stability of the laminar air flow is lost and sterility can no longer be maintained. The pressure drop can be monitored with a manometer fitted to the cabinet, but direct measurement or airflow with an anemometer is preferable.

Routine maintenance checks are required (every 3-6 months) of the primary filters, which may be removed (after switching off the fan) And washed in soap and water, as they are usually made of polyurethane foam. Every 6 months the main filter should be checked for air flow and holes (detectable by locally increased air flow and an increased particulate count). This is best done on a contract basis.

Regular weekly checks should be made below the work surface and any spillage mopped up and the area sterilized. Spillages should,

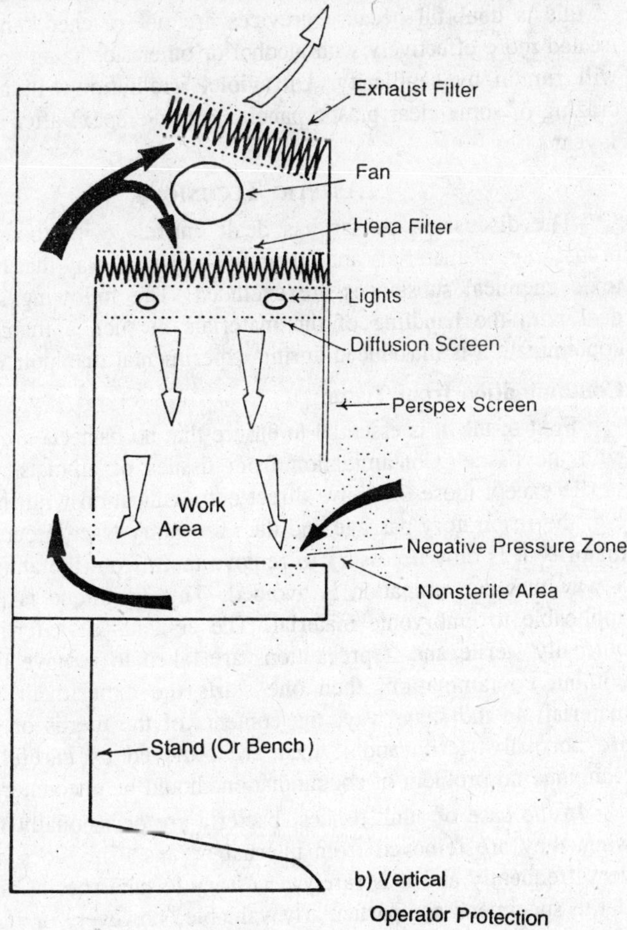

Fig. 8.3. Vertical laminar flow hoods. Filled arrows, nonsterile air; open arrows, sterile air.

of course, be mopped up when they occur; but occasionally they go unnoticed, so a regular check is imperative.

Laminar flow hoods are best left running continuously because this keeps the working area clean. Should any spillage occur, either on the filter or below the work surface, it dries fairly rapidly in sterile air, reducing the chance of growth of microorganisms.

Ultraviolet (uv) lights are used to sterilize the air and exposed work surfaces in laminar flow cabinets between use. The effectiveness of this is doubtful because crevices are not reached, and these are treated more effectively with alcohol or other sterilizing agents, which will run in by capillarity. Ultraviolet irradiation will also lead to crazing of some clear plastic panels (e.g., Perspex) after 6 months to 1 year.

Aseptic Technique

The discussion so far has dealt entirely with the preliminary preparation of materials and apparatus in such a way that bacteria and toxic chemical substances are excluded. The following section will deal with the handling of the materials in such a manner that no contamination is introduced during experimental manipulations.

Contamination from Tissue

First of all, it is essential to ensure that no bacteria are introduced with the tissue. Contamination from tissues of animals are already sterile except those that have direct communication with the exterior, e.g. the respiratory passage and the alimentary tract. Hence, with the majority it is only necessary to remove them from the animal in such a way that contamination is avoided. This technique is particularly applicable to embryonic material. The contents of a fertile egg are normally sterile and if precautions are taken to remove the embryo without contamination, then one starts the experiment with clean material. In the same way, the contents of the uterus of a mammal are normally sterile and if they are removed by careful operating technique no problem of contamination should be encountered.

In the case of adult tissues, bacteria are occasionally found even when they are removed from internal organs. This does not happen very frequently and it is rarely necessary to take special precautions. If the specimens are particularly valuable, however, or if there is a risk that they may have been contaminated during removal, as is often the case with surgical specimens before they have reached the tissue culture laboratory, then they must be sterilized. This can be done effectively by means of antibiotics, the tissue being washed in a solution of very high concentration before being grown in a medium containing the usual concentrations. A useful antibiotic mixture is balanced salt solution containing 1,000 units of *penicillin* per ml. and 0.5 mg. Of streptomycin per ml. *Neomycin* (0.5 mg. Per ml.) is also highly effective. If the tissues are washed in this solution before they are cut up for explanation, very few organisms will survive.

The elimination of microorganisms is a particularly big problem in growing tissues from cold-blooded organisms and invertebrate. However, the development of antibiotics has greatly facilitated this and a procedure such as that described for surgical specimens is suitable. Other means of sterilization have also been used, e.g. 1 in 1,000 merthiolate, ultraviolct light and 10 per cent hexylresorcinol.

Contamination from the Air

If no mistakes in technique have been made in preparation of the tissue, media and apparatus, then subsequent contamination can only arise from the air in the room in which the operations are carried out, or from the operator. Contamination from these sources can be avoided in almost all cases by adhering to a few simple rules of aseptic technique. Only rarely it is necessary to use a specially fitted room and the desirability of doing so depends on the problem to be undertaken. For instance, many experimental programs involve the use of short-term cultures, that is to say, cultures grown for a period of some days only. In these cases, it is unnecessary to take special precautions since with careful technique it is perfectly possible to carry out short-term tissue culture in an ordinary laboratory. When this is done there may be occasional losses due to contamination by dust or spores from the air, but these cultures can be discarded with no great loss of time or results.

On the other hand, moderately long-term experiments, i.e. experiments lasting two to three weeks, make an aseptic room desirable. This room need not be specially fitted out. It need only be a room kept apart for carrying out tissue culture procedures. It should be kept particularly clean and only operators should be permitted in the room while tissue cultures are being handled. If this is done and careful aseptic technique employed, other precautions are not necessary.

When an aseptic room is regularly used by a number of people, it is desirable to have a set of rules concerning its use. The following set of rules has been found adequate.

1. The room must be kept clean and tidy at all times.
2. No materials known to be contaminated may be opened in it.
3. When aseptic work is being performed nobody may enter the room or open the door without obtaining permission from the person working there.
4. All working areas should be swabbed with alcohol before use.
5. If discarded medium is placed in the sink it must be washed down thoroughly with water for several minutes.
6. After use the table must be cleared and cleaned. All materials must be removed. All marks due to drops of medium etc. must be washed off.
7. Pipettes and test-tubes must be removed from the rack immediately after use. Unused materials must be removed for resterilization.
8. Used pipettes must be placed in the jar of water. Other glassware which has been in contact with medium or other proteinaceous solutions should be placed in the pail of water. The last person to use the room each day must ensure that the jar and pail are removed to the washing up bench. Each morning the jar and pail must be refilled with water and replaced in the aseptic room first thing.
9. Clean, used glassware should be collected in a wire basket and removed to the washing up bench.
10. Only articles actually belonging in the aseptic room may be left there.

Long-term experiments with very valuable materials may require more rigorous precautions. Thus it may be desirable to sterilize the air in the aseptic room. The normal practice is to circulate it through a series of filters but it may also be passed through an ultraviolet irradiation chamber where all organisms are killed while a further refinement, sometimes introduced, is an electrostatic dust remover. In the absence of a circulating system, the air can be partially sterilized by keeping an ultraviolet on continuously. It must be so situated that it does not play on the operators who might otherwise suffer injury. This can be done by mounting it above head level in such a position that all radiation is directed upwards. In this way all the air above the light is sterilized, so that no bacteria-laden dust can fall in the operation zone.

Further precautions that should be taken in operating an aseptic room are aimed at the laying of dust. Floors must be carefully cleaned at regular intervals and both they and the walls should be treated with a light oil which will cause dust to adhere. In the author's opinion many of these precautions are unnecessary except where operators are unskilled or where the material is particularly valuable or likely to be dangerous if any contamination at all occurs, e.g., in vaccine production.

Contamination from the Operator

If all the preceding sources of contamination have been eliminated then the one source left is the operator himself. Contamination as a result of the presence of an operator can arise from three sources:

1. Direct contamination of instruments or cultures from non-sterile objects during operation.
2. Bacteria carried in expired air.
3. Organisms carried by dust raised by the operator's movements.

Direct contamination may arise from:

1. Contact of apparatus, particularly pipettes, with non-sterile surroundings or with the hands.
2. The entry of organisms present at the lip of containers.

For the beginner, the most difficult part of aseptic technique is the mastery of various manipulations without contaminating pipettes by bringing them into contact with non-sterile objects. The only way to learn to avoid this is to perform the manipulations repeatedly under the supervision of an experienced technician who will point out each fault. During the handling or cultures, if there should ever be any doubt as to whether a pipette touched a non-sterile object; it should invariably be discarded. Contamination from the hands may be minimized by sterilizing them before commencing work. This can be done fairly efficiently by through washing in soap and water. Some people prefer to rinse them with a little alcohol.

One particular point to watch is *double transfer* of contaminating organisms, i.e., contamination of a culture with an instrument which has itself been contaminated by contact with the hands. There are two very common examples of this. A frequent mistake is to hold a pipette too low down while fixing a rubber tube or teat to the end. If this pipette is subsequently inserted into a bottle or container, it will contaminate the inside of the neck. When a pipette is again inserted, bacteria may be carried right into the medium. A second example of this type of contamination is that commonly encountered in the handling of knives. These instruments are usually grasped well forward when used. If they are then placed inside sterile tubes or other containers and subsequently withdrawn, the tips the blades will be contaminated as they pass over the area which has been in contact with the contaminated part of the knife handles.

These errors can be avoided by practice and meticulous attention to technique, which ultimately becomes quite automatic. There are, in addition, a number of simple routine precautions that can taken to avoid some of the errors described. Thus all pipettes should be discarded immediately after use. This avoids the possibility of contamination due to using pipettes which have been rendered non-sterile by being placed in contaminated holders and is well worth the extra outlay and labour. Also it is often desirable to use a system for instruments which involves their re-sterilization immediately before use. Two methods are commonly used. In one, the instruments are placed in a boiling water bath after use and removed again just before they are required. In the other technique, the instruments are dipped in spirit and this is burned off before they are re-used.

On opening culture vessels there is always a great danger of contaminating the contents with organisms from the lip. This happens because the vessels accumulate a little dust in the incubator. Also on cooling to room temperature a negative pressure is created within them so that when the stopper is removed, there is an inrush or air which carries the dust with it. This can largely be avoided by flaming the necks of all vessels before removing the stoppers. The flaming need only brief since its function is simply to fix the dust in place but it should be noted that all parts of the neck of the vessel must be flamed. It is not sufficient just to place the neck briefly in the flame. It should be rotated while it is there.

If the medium comes in contact with the neck of the vessel, as may happen if the contents are removed by pouring, then this cursory type of flaming is not enough since bacteria will be washed back into the flask with the last drop or two remaining. In such cases, the neck of the vessel must be very carefully flamed till it is not enough to kill any bacteria present. The contamination of cultures by dust from the lip of the vessel can be minimized by covering the stopper and neck. A brown paper cover secured with a rubber band is adequate or alternatively an aluminum foil or parafilm cover can be used. Screw-caps have an advantage over stoppers in this respect since they cover a small area of the neck.

Many people regard contamination from the breath of the operator as one of the main causes of infection of cultures. In fact, this is not so. During ordinary quiet breathing very few bacteria are exhaled. The number increases very greatly, however during speed or coughing. Therefore, contamination from exhaled bacteria can be prevented in one or two ways, either by using a mask or by refraining from taking. There is some danger in using a mask in as much as it tends to create a sense of false security. A mask is quite efficient so long as the operator is breathing quietly but if he speaks or coughs it becomes filled with organisms and subsequently even quiet breathing forces them out of the cloth. Instead of a thick cloth mask, it may therefore be preferable to use a baffle made of X-ray film. Alternatively, a glass plate may be interposed between the operator and the material on which he is operating. This is particularly desirable if conditions make it necessary for him to lean over his work. Otherwise, his problem is best handled by refraining from taking and by keeping the mouth and nose well away from the cultures.

Contamination from dust can be avoided by (a) preventing dust from being raised in the room, and (b) preventing dust in the air from entering the culture vessel.

Dust normally arises from three sources:

1. The floor and operating table.
2. The clothes.
3. The operator's hair and body.

The raising of dust from the floor and table can be prevented by coating them with a layer of light oil. Dust adheres to this and is prevented from spreading about the room. Sometimes it is undesirable to oil the table top and is that case it can be swabbed with 70 per cent alcohol before use to kill off most of the organisms present. Even when these precautions have been taken, it is highly desirable to avoid all unnecessary movement in the room and all draughts should be excluded.

Clothes, particularly street clothes, carry a great deal of dust and any movement within the room is likely to set up a cloud of germ-laden particles. To a large extent this can be prevented by wearing a clean laboratory coat over the street clothes. For particularly careful work a clean, sterile gown should be used on each occasion, although for ordinary work this is not necessary. If a laboratory coat is not worn, at least the sleeves should be rolled up and the arms washed up to the elbows. While working with cultures it should always be assumed that dust will fall from the arms so that care should be taken to avoid passing the hands and arms over open dishes.

One of the most dangerous sources of dust is the operator's hair. If the procedures employed necessitate bringing the head over the working table or near it, it is better to wear a cap. Also, the operator should try to keep as far away from his work as is convenient.

Despite all these precautions, it is still possible that there may be some dust in the air and one must, therefore, continuously employ a technique which will prevent it from settling into the culture vessels. This can be achieved by keeping all vessels covered as much as possible and it is often desirable to perform all manipulations under a glass plate. When bottles and flasks are handled they should be held at an angle when open so that dust cannot drop straight down into the bottom.

9

Culture Media

Natural Media

The medium is by far the most important single factor in culturing cells and tissues. Ideally, this medium should be an accurately defined mixture of chemical substances. For the culture of plant cells this ideal was achieved many years ago, but at the time of writing we are only approaching it for animal cells, although already there are one or two synthetic mixtures which will support indefinitely the growth of one or two special strains of cells.

On the whole, animal cells need either a completely natural medium or a medium supplemented with some natural product. Although improved synthetic media will be produced in increasing numbers in the next few year, it seems certain that natural media will form the cheapest and most convenient materials for the maintenance of cells for some time to come. Also, it is highly likely that natural media will remain essential for growing newly isolated tissues and highly demanding cells for very many years.

The natural materials used to promote cell growth fall into three general categories: coagula, such as plasma clots; biological fluids, such as serum, and tissue extracts, of which the commonest types is embryo extract

Plasma

In the traditional methods of tissue culture, the tissue fragment is grown in a coagulum. In his original experiments, Ross Harrison grew a small piece of tissue from near the neural tube in a clot of frog lymph. Later, Carrel and Burrows introduced plasma for the same purpose and plasma clots have been in general use since. Although most of the newer techniques of tissue culture do not employ a plasma clot and the vast majority of cells are now grown without one, some tissues can only be grown in the presence of a supporting matrix. The usual matrix in use is *fowl plasma*. It is now available commercially from several sources and can be bought either as *liquid plasma* in siliconed ampoules or as *lyophilized plasma*, to be reconstituted by the addition of distilled water saturated with carbon dioxide.

Some people prefer to prepare their own plasma and in certain circumstances it is more suitable to do so. Any animal can provide plasma, but the fowl is generally used since fowl plasma gives a more solid clot than most mammalian plasmas and, in addition, is easily obtained. The male bird is usually employed because the blood calcium level is more constant and in certain procedure such as bleeding from the carotid artery, it is easier to operate on the male. A young bird is preferred and it is usually stated that it should be less than one year old.

When collecting plasma, it is essential to prevent coagulation of blood. This can be done either by preventing the initiation of coagulation or by interfering with the coagulation process. Coagulation is initiated either by contact of the plasma with a water-wettable glass surface or by the release of products of tissue damage into the blood. By avoiding tissue damage and by coating the glass vessel with a water-repellent material, coagulation can be prevented without the addition of anticoagulant. In most circumstances, however, it is impossible to prevent tissue damage and coagulation is therefore retarded by one of several methods. Heparin is the most commonly used anticoagulant. It has been shown that in high concentrations it can be inhibitory to the growth of tissue cells and may produce morphological abnormalities, but if the concentration of heparin is kept at such a level that it just inhibits coagulation (0-2 mg. Per cent) it seems to be relatively harmless. The onset of coagulation may also be inhibited by cooling and this is very commonly used as an additional measure, the blood being collected in tubes cooled in ice. In blood transfusion practice, coagulation is commonly prevented by inhibiting the ionization of calcium by the addition of sodium citrate to the blood, but this is not employed in collecting plasma for tissue culture purposes.

There are three methods of removing blood conveniently from bird—*wing bleeding*, *heart bleeding* from the carotid artery. Where there is no objection to the addition of an anticoagulant to the blood, wing or heart bleeding may be employed. However, when it is desirable to avoid the addition of an anticoagulant it is advisable to resort to one of the more complicated procedures such as *carotid bleeding* where the contamination of blood by tissue material can be kept to a minimum.

Bleeding From the Wing

Bleeding from the wing of a young bird requires a certain amount of knack but when this is learned it is fairly easy procedure. It is advisable to treat the syringe, needle and tubes for blood with heparin. This is added to the tubes as a 25 mg per cent solution in amounts such that the final concentration will be 0.2 mg per cent in the blood. It is also advisable to cool them in ice while the blood is being collected. A 20 ml. syringe may be used with an 18 or 19 gauge needle.

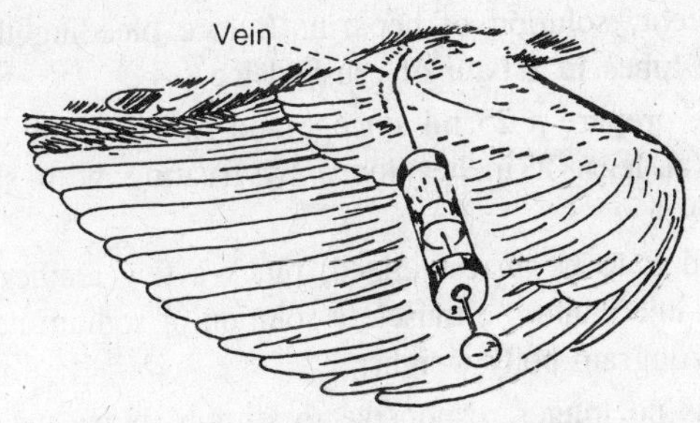

Fig. 9.1. Bleeding from the wing.

The bird is placed on its side or back and wing is fanned out with the underside uppermost. The feathers are then plucked from the region of the 'elbow'. On clearing this area a large vein can be seen to pass across it, as shown in the diagram. The area is swabbled with alcohol and when this has evaporated, the needle is inserted. This should be done with great care as it is very easy to go right through the vein and cause a large haematoma which will make subsequent exploration futile. When the vein is entered blood can be withdrawn gently. If too much suction is applied, the wall of the vein may be sucked into the lumen of the needle and the flow of blood will stop. When sufficient blood has been collected in this way, it is transferred to a prepared centrifuge tube and spun down as soon as possible.

Bleeding from the Heart

As a routine procedure, heart puncture is probably the easiest method of obtaining a fairly large quantity of heparinized blood from a fowl and it will therefore be described in detail.

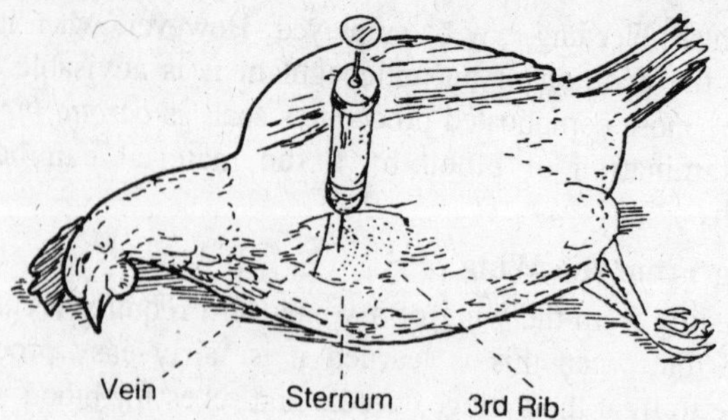

Fig. 9.2. Heart puncture; first method.

1. Prepare a number of centrifuge tubes by coating by coating them with silicone (Repelcote, Hopkins and Williams). Stopper with cotton wool and sterilize by dry heat. To each add 0-1 ml of a 25 mg per cent, solution of heparin. Before bleeding the cockerel place the tubes in a bath of crushed ice.
2. Similarly, prepare a 25 ml syringe and two or three No.2 serum needles (at least 2 inches long) by treating with silicone and sterilizing.
3. If the bird is to be anaesthetized, pluck a few feathers from one thigh and inject into the muscle a solution of sodium nembutal (25 mg. per kilogram body weight).
4. When it is no longer responsive to stimuli, place the bird on its right side. It can be secured to the board by a tape round the neck and another round the legs and the wings may be crossed over each other to prevent flapping. Instead of securing the bird to a board, it can be steadied by an assistant who should face the operator, grasping the legs in his left hand and restraining the wings with his right hand while the head is held under his right elbow.
5. Pluck the small feathers over the cardiac region and press the larger ones aside. The first landmark to be seen is a large vein running dorsally. Behind this a soft area can be left an the dorsal edge of the breast muscle can be located crossing it. Swab the skin with alcohol and insert the needle at this point. If it strikes the sternum withdraw it partially and re-enter slightly further back. The angle of syringe and needle should be almost vertical. When the needle has been inserted one to one and a half inches the pulsation of the heart should be felt. Advance the needle a little further, applying gentle suction until blood enters it, then aspirate fairly rapidly.

Fig. 9.3. Heart puncture; second method.

6. Immediately transfer the blood to a cooled tube and centrifuge (5-10 minutes at 1,800 g, 2,500 r.p.m. in the ordinary clinical type centrifuge). After centrifugation transfer the plasma layer to clean, sterile, siliconed tubes and store in the refrigerator. Take a sample for sterility testing at the same time. (When centrifuging, the cotton-wool plugs must be fastened in position by turning some of the cotton over the lip of the tubes and securing with an elastic band).

Note : In order to obtain clear plasma, the bird should be deprived of flood for at least eight hours before bleeding. It should, however, receive a plentiful supply of drinking water. After bleeding it can be fed once more and must have plenty of water since it will be very thirsty. It is unusual to remove more than 25 ml of blood at a time by this method but up to 75 ml can be taken if desired. A practised operator can dispense with heparin but unless the bleeding can be carried out with virtually no tissue damage its use is recommended.

An alternative and possibly easier way to bleed a chicken is to place the bird on its back with its head lying the end of the table. The area at the base of the neck is then swabbed with alcohol and the small feathers plucked as before. A syringe with a 2-inch needle is inserted just behind the sternum and passed back horizontally until it enters the heart. Blood can then be withdrawn as required.

Bleeding from the Carotid Artery

It is desired to remove blood without the addition of anticoagulant, it is usually necessary to bleed from the carotid artery. This requires

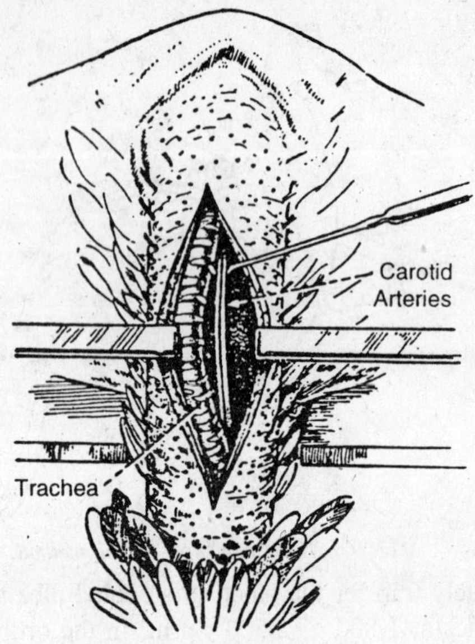

Fig. 9.4. General anatomical relationships of the carotid arteries in the fowl.

an operation which is quite simple although it needs some practice. The bird is placed on its back on an operating board with its head failing over the end of the table. It is anaesthized either with ether or by injection of sodium nembutal into the thigh muscle some time in advance (25 mg/kilo body weight). An incision is made down the middle line of the throat and a raphe is then seen in the midline. The muscles are separated from this on either side and underneath them the two carotid arteries are found running side by side. A curved needle is inserted behind these to clear them from the surrounding tissues. One artery is freed from the other and a double ligature is passed under it. The distal ligature is tied. Now one of two procedures may be adopted. In the first, some sterile olive oil or liquid paraffin is poured over the area and a small incision made in the wall of the artery (without cutting right through). When this incision is made, blood begins to spurt forth and may be collected in prepared siliconed tubes. This procedure is satisfactory, but many people prefer to insert a sliver cannula into the proximal end of the artery after making an incision and to direct blood into the siliconed centrifuge tubes by this means. The bird may be bled to death or the artery ligated and the skin incision repaired. It will then be able to provide another supply of blood in a few weeks from the other artery.

Collagen

There is little doubt that many newly isolated tissues grow better on a physiological substrate such as plasma. Unfortunately fibrin substrates tend to liquefy. Fairly recently, Ehrmann and Gey suggested the use reconstituted rat-tail collagen. It has proved particularly useful for some morphological studies and has been adapted for both test-tube and coverslip cultures.

Rat-tails are sterilized in 95 per cent, alcohol and then fractured, the broken pieces being removed to expose the tendons. The tendons from one tail are submerged in 150 ml of 1:1000 acetic acid solution and left, sealed, in the refrigerator for 48 hours. Prolonged centrifugation (1 hour at 2500 g.) is then required to separate the residue from the acetic acid solution of collagen. A further two extractions may be performed. The material can be stored for some months in the refrigerator. Before use it is dialyzed against several changes of distilled water. During dialysis the viscosity of the solution increases. Ultimately it becomes too viscous to handle and dialysis should be stopped before this stage, according to whether a thin or thick collagen layer is required. Usually the minimum time of dialysis is 24 hours and the maximum 48 hours. For use the solution is spread over the surface of the vessel and collagen precipitated as a thin tough film by exposure to ammonia vapour. Finally the ammonia is removed by washing with distilled water and then balanced salt solution. The tissue is grown direct on the collagen, as on a glass surface.

Biological Fluids

Despite the many advances in our knowledge of the nutrition of tissue cultures in the past few years, the basis of most media is still one or other of a number of biological fluids. The traditional biological fluid and the one most commonly used in serum. There seems to be no advantage in using autologous or homologous serum except possibly in the case of a few adult tissues. A percentage of sera are always toxic but autologous sera may be just as toxic as heterologous ones and the rule is that a serum is either toxic or non-toxic irrespective of its origin. Serum is most commonly obtained from human adult or placental cord blood, horse blood, or calf blood.

Preparation of Serum

The preparation of serum is a straight forward procedure. It consists of permitting whole blood or plasma to coagulate and sinerese and thereafter removing the exuded serum. Serum thus collected has to be tested for sterility before it is used and it is also advisable to filter it. While waiting for the result of the sterility test, it is, of course, necessary to store the serum and it is recommended that this should be done at low temperature, preferably in deep freeze. Opinions differ as to the value of storing serum in very cold conditions but in view of the established importance of glutamine as a nutrient and its known instability, it would seem desirable to freeze serum when it is to be kept for any length of time.

It is recommended that the serum should be tested for toxicity and this is best done by making up large numbers of small cultures and testing them with a medium containing the unknown serum before using it for experiments. For this purpose one can use either primary explants or, more conveniently, small replicate cultures of a standard cell strain such as the 'L' cell or HeLa cell. The toxicity of sera can also be avoided to some extent by using pooled material.

Placental Cord Serum

Placental cord serum is strongly recommended by some authorities, in particular Dr. Gey and Dr. Margaret Murray. If possible, the attendants should have been requested not to strip the cord at the time of birth since this is apt to cause haemolysis. The blood is allowed to drip from the umbilical cord into a receiving vessel. The placental blood thus collected is allowed to clot and the serum is pipetted off. It is usually necessary to centrifuge it in order to remove red blood cells. This serum may be filtered but a common practice is to leave it at temperature for two or three weeks during with time it tends to be self-sterilizing. If any bacterial growth occurs during this time it should be rejected. This serum is considered to be particularly good for the growth of tissues which are otherwise very difficult to culture and is said never to be toxic.

Amniotic Fluid

Amniotic fluid from various sources has been used in many tissue culture media. The most conveniently obtained amniotic fluid is bovine. A pregnant uterus is obtained from the local slaughter house. It is carefully washed on the outside and then hung from a hook. An area near the base is swabbed with iodine and after this has dried a trocar and cannula with tube attached is inserted into the uterus and the amniotic fluid allowed to drain off into suitable vessels. After sterility testing this material is ready for use.

Ascitic and Pleural Fluid

Ascitic and pleural fluid have both been described as useful culture media and ascitic fluid from patients with peritoneal carcinomatosis is particularly recommended. As a general rule these materials will be obtained from a hospital unit and they can be used as soon as their sterility has been verified. It is advisable to test them for toxicity since many of the patients producing these fluids are receiving heavy medication and some of the drugs may be present in toxic concentrations in the fluids.

Aqueous Humour

Aqueous humour from the eye has been suggested as a good nutrient for cultures since in vivo it is the sole source of nutrients for the corneal epithelium. The best source of aqueous humour is the ox eye. These can be obtained in larger quantities form the local slaughter-house but of course the collection of sizeable amounts is tedious and probably aqueous humour is desirable only for rather small-scale work.

Serum Ultrafiltrates

It has been established fairly satisfactorily that the most important components of serum are the large molecules, that is to say, the proteins and protein-bound materials. These are apparently irreplaceable except in a very few instances. However, the ultrafiltrate of serum is a very useful supplementary nutrient. It was developed by Simms to prevent the accumulation of fat in tissue during cultivation. Its preparation requires special apparatus and is somewhat tedious but serum ultrafiltrate is available commercially and may be found a usefui supplement in growing some highly demanding tissues.

Tissue Extracts

Almost all media used for the cultivation of tissues until very recently contained some sort of tissue extract, usually embryo extract. In the past few years, effective substitutes for embryo extracts have been developed so that it iess used than previously, and it seems certain that its use will diminish progressively. It has, of course, been known for many years that cells will grow without embryo extract and some tissues have been cultivated quite successfully in serum and balanced salt solution alone. It has been found that the most important components of embryo extract are in the small molecular fraction and that they can be effectively replaced by synthetic supplements which are usually mixtures of amino-acids.

There are, however, exceptions and, especially in the case of differentiated tissue, there may be factors in embryo extracts which can not be adequately replaced. Harris and Kutsky have demonstrated that, while the small molecular fraction constitutes the essential component of embryo extract, there is also a nucleoprotein fraction which greatly improves the growth of newly isolated cells.

The Preparation of Embryo Extract

Broadly speaking there are two different kinds of extract. There are those prepared with the minimum of damage to the issue and others which involve complete destruction and are really homogenates. These two extracts produce quite different effects on cells. For instance, the homogenate type has to be used with care since in high concentrations it can be quite toxic.

Preparation of Chick Embryo Extract

This preparation provides an example of a simple extract from relatively intact cells. Embryos of different ages may be used. If they are young, that is to say incubated up to ten days, a simple embryo juice may be prepared by placing two or three of them in a barrel of a sterile 20 ml. syringe and expressing them into a suitable receiver. This juice is diluted with balanced salt solution of 50 per cent and after centrifugation the supernatant constitutes EE_{50}. The procedure is detailed here.

The Preparation of Bovine Embryo Extract

This, on the other hand, provides an example of the homogenate type of extract. A pregnant uterus is obtained from the slaughter-house and the foetus is palpated. A uterus should be selected which appears to contain an embryo from four to six inches long. The outside is carefully washed and swabbed with iodine. It is best first to remove the amniotic fluid by means of trocar and cannula. After this the uterus can be placed in a large sterile tray. A long incision is them made and the wall folded back. The embryo presents itself and can be removed quite easily. It is chopped into fairly small pieces and placed in a Waring blender with approximately the same volume of balanced salt solution. After homogenizing a high speed for two to five minutes the resulting pulp is transferred to centrifuge bottles and spun down. The supernatant fluid, which is usually cloudy in this case, constitutes EE50. Brayant, Earle and Peppers have developed a considerable refinement of this technique. They suggest that the product described should be treated with hyaluronidase and then ultracentrifuged in the Spinco to remove all solid matter. Thereafter it can be filtered through a sintered glass filter and the filtrate provides an excellent growth-promoting substance.

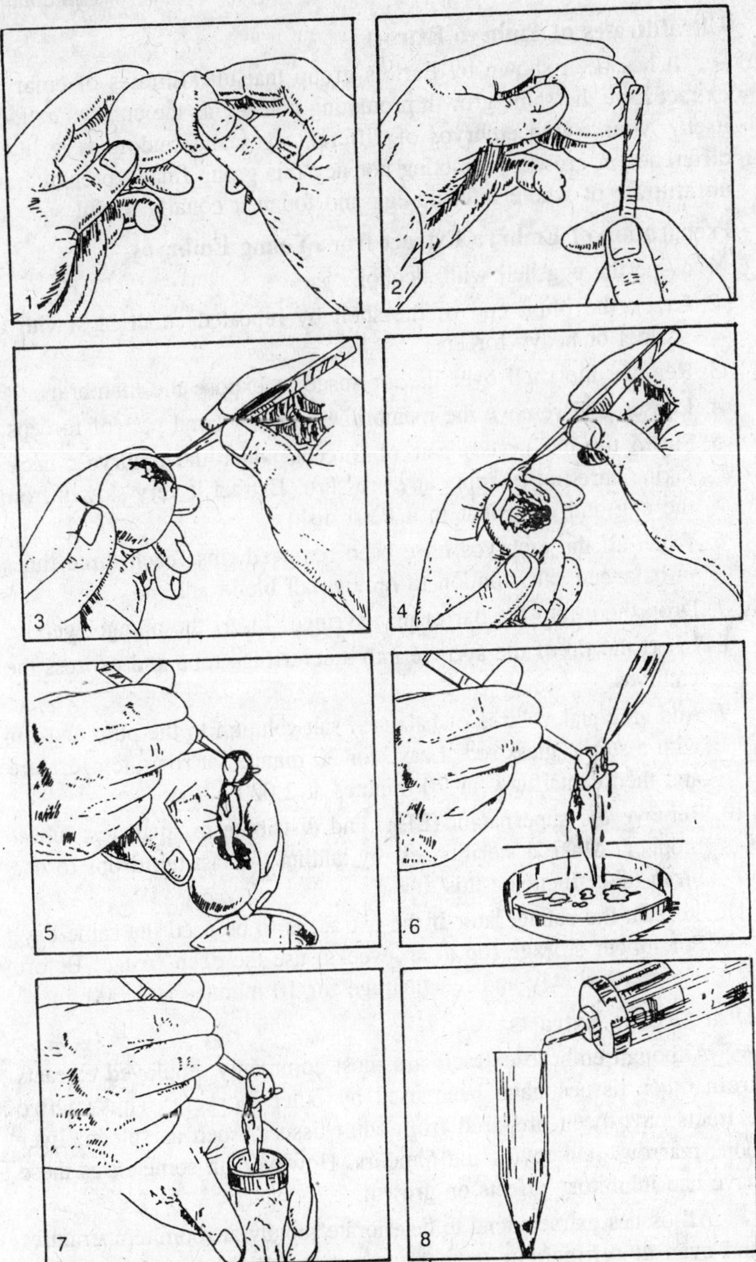

Fig. 9.5. Preparation of embryo extract.

Ultrafiltrates of Embryo Extract

It has been shown by Earle's group that ultrafiltrates of embryo extract have the same growth promoting potency as the embryo extract itself. After trying embryos of different ages and finding very litter difference in growth promoting potency this group finally prepared an ultrafiltrate of whole infertile egg and found it equally useful.

Preparation of Embryo Extract from Young Embryos

1. Wipe the eggshell with alcohol.
2. Crack the blunt end of the shell by repeated small taps with a scalpel or heavy forceps.
3. Remove the shell over the air space to expose the membrane.
4. Loosen and remove the membrane with a second pair of forceps.
5. Slip a third (Curved) pair of forceps under the embryo's neck, taking care not to apply any pressure. Extract it very slowly from the egg and deposit it in a Petri dish.
6. When all the embryos have been removed rinse each three times in balanced salt solution to remove all blood and yolk.
7. Drop them into the barrel of a syringe. Insert the plunger gently.
8. Insert the tip of the syringe into a centrifuge tube and express the embryos.
9. Add an equal volume of balanced salt solution to the pulp and stir with a sterile glass rod. Leave for 30 minutes at room temperature and then centrifuge for 20 minutes at 2,000 g.
10. Remove the supernatant (EE_{50}) and distribute to small test tubes. Stopper. Make a sterility test by adding the last ten drops to one broth tube. Incubate this tube.
11. Store in the refrigerator if the extract is to be used the same day. For longer storage (up to six weeks) use the deep freeze. Before use, thaw slowly and recentrifuge for 10 minutes at 2,000 g.

Other Tissue Extracts

Although embryo extracts are most commonly employed extracts from other tissues have been used by other workers. Thus, active extracts have been prepared from adult tissues, such as spleen, liver bone marrow, leukocytes and tumours. However, in some cases these have had inhibitory effects on growth.

All tissues extracts tend to deteriorate rapidly at room temperatures and even at refrigerator temperatures they will lose a great deal of their growth-promoting potency within a week. They can, however, be stored satisfactorily at 20° and it is best to store them in this way in small quantities. On thawing the extracts a precipitate will be found to have formed and it is necessary therefore to recentrifuge immediately before use.

Other Media of Biological Origin

Since the major nutritional requirements of cultured cells are likely to resemble those of many micro-organisms, it is natural that attempts have been made to employ bacteriological media for their growth. These attempts have met with some success though most bacteriological media are inadequate for prolonged maintenance of animal cells.

The most useful supplement of this sort is lactalbumin hydrolysate. For many cell types a medium consisting of 20 percent, serum and 80 per cent, balanced salt solution, with the addition of 0.5 percent, lactalbumin hydrolysate, proves adequate. Since this medium is easily prepared and handled it is particularly convenient one for use in the maintenance of cell stocks. It is not adequate for many 'fibroblastic' strains.

Other supplements, such as yeast extract, tryptic meat broth and Filde's medium have been found partly useful.

Defined Media

While tissue culture media obtained from natural sources are still widely used for many purpose, especially for growing tissue freshly isolated from the organism, they suffer from obvious disadvantages. In particular, since their composition is unknown and variable, it is virtually impossible to reproduce conditions exactly from one experiment to another. For this reasons attempts have been made from the very earliest days to develop media composed entirely of defined components. The problem was tackled in two different ways—the so-called 'analytical' and 'synthetic' approaches. In the first, attempts were made to analyse the medium and identify essential components of the intact animal, were combined to form media which were tested for their ability to keep cells alive. It is readily apparent that the division is an artificial one since the two methods of approach are mutually dependent.

With our present knowledge of metabolism it is obvious that there will never be a universal medium for all cell-types. Thus, the requirements for plant cells and animal cells have many things in common but are, on the whole, quite different. Among animals, it is not surprising that the requirements for insect tissues should be different from those for mammalian tissues. And although most media for mammalian cells seem to be equally effective for different cell types, it must not be forgotten that different absolute requirements exist between different animals. For instance, ascorbic acid is an essential dietary constituent for man, the higher apes and the guinea pig, but it is not required by other mammals and one might expect such differences to be reflected at some level in the cells. Within the same organism it would also prove surprising if all cells had precisely the same requirements. Thus, cells with strong proteases might be able to use intact protein while cells with no proteases would require the constituent amino-acids. Also for full function we would expect specialized tissues such as thyroid epithelium to have special requirements (in this case iodine and rather large quantities of tyrosine). While these possibilities must be kept in min, it has to be realized that at the present moment we are at an early stage in the development of special media and most recent achievement have been in producing media of general use.

A substance added to a medium may be beneficial, harmless or harmful. If the effect is beneficial it may be because the substance is essential for the cells or because it has a supplementary value, and it is important to distinguish between these. If it is harmless it may be because it does not enter into the metabolism of the cell or because it is present in an insufficient amount to produce an effect. It is harmful it may be because it is intrinsically poisonous to the cell or because it is present in an unsuitable concentration. Very few studies have been done to define the precise value, in quantitative terms, of the metabolites added to defined media for mammalian cells.

Most media consist of mixtures of substances which have proved essential, generally beneficial or harmless. Keeping in mind that our knowledge is incomplete and that most 'synthetic media' have been derived empirically, we will now consider the different types, in relation to the culture of mammalian cells.

Media can conveniently be classified in four groups:

1. Media essential for immediate survival.
2. Media essential for prolonged survival.
3. Media essential for indefinite growth.
4. Media essential for specialized functions.

The components essential for the immediate survival of cells and tissues have been precisely defined. It is essential to control osmotic pressure and pH, while a source of energy and certain inorganic ions are necessary for all but the briefest survival. These requirements are met by a combination of salts and glucose. A simple medium of this sort is referred to as a *balanced salt solution*. These salt solutions are the basis of all culture media except the very simplest of natural media.

Cells and tissues will survive for a brief time in balanced salt solution but for more prolonged survival other factors are necessary. These factors, normally present in whole serum, can be replaced by a synthetic mixture. For survival for long periods of time mammalian cells require, in addition to a balanced salt solution, all the essential amino-acids, oxygen, vitamins and serum protein. If the serum protein is not exhaustively dialyzed, this mixture may support good growth almost indefinitely but mixtures containing exhaustively dialyzed serum will not do this as a rule. In the complete absence of protein the cells do not survive long except in a very few instances of cells which have been specially adapted. The best example of such a medium is the one developed by Eagle.

Recent years have witnessed rapid progress in the development of entirely synthetic media. At present, however, very few cell strains can be grown in media from which serum has been entirely omitted and greatest success has been achieved with the strain L cell. Medium 1066, developed by Parker and his colleagues and medium NCTC 109 developed in Earle's laboratory were the first media in which prolonged multiplication of the strain L cell was observed. These media resemble each other closely and are distinguished by their complexity. Some of the components used, especially certain of the co-enzymes are relatively impure and therefore some unknown components are almost certainly present. Waymouth's medium 752/1 represents a considerable advance since it is very much simpler than the other two and contains fewer components of dubious purity HERT 1 medium is capable of maintaining indefinite survival of a subline of strain L cells and is almost as simple as Eagle's medium. It should be reemphasized that at the present time most other cell-strain will fail to survive in media containing no serum. Some intermediary media, containing peptone are, however, capable of supporting growth of some of these more exacting strains.

The fourth type of medium—that essential for specialized function—has hardly reached the earliest stage of development. It seems likely that such media are best based on one of the complex defined media mentioned in the previous paragraph. Requirements for each organ will have to be defined as a result of planned investigations. Certain facts are know. Among the best established is that vitamin A is an essential component of media for the growth of ciliated epithelia.

Similarly, the survival of many tissues is dependent on the presence of appropriate hormones. For example, adrenal cortex will survive only in the proximity of pituitary tissue and probably requires adreno-corticotrophic hormone. Also some female sex organs are dependent on oestrogens in the medium. Many such requirements can be deduced from a knowledge of the behaviour of the tissues *in vitro*. However, the application of this knowledge to the maintenance of cells and tissues *in vitro* in defined media is still at such an early stage that no medium of this sort has been established.

Before proceeding to consider special examples of the above classes of media some chemical factors will be mentioned which must be taken into account in preparing them.

Solubility of Materials

Many chemicals used in synthetic media are very sparingly soluble and it is essential to prepare them in such a way that they will come into solution. Thus tyrosine is only soluble in acid solution and even then only to a limited extent. Therefore it has to made up in a special acid stock solution and mixed in with the other materials at a late stage in preparation when the addition of acid will do no harm. Lipids are particularly difficult to dissolve and they have to be taken up in alcoholic solutions. These have to be sufficiently concentrated that the alcohol will be greatly diluted when added to the rest of the medium. Also a non-toxic detergent (Tween 80) has to be added to stabilize the solution. One of the most elementary difficulties concerns calcium phosphate which is highly insoluble, particularly in alkaline solutions. Therefore, in making up salt solutions it is essential that calcium and phosphate should not meet until the solution is quite dilute and that it should not be made alkaline until the last minute.

Compatibility of Components

In combining certain materials it must be remembered that they may react together to produce quite different substances. This does not readily happen at incubator temperature but may occur on heating solutions. Thus, if ascorbic acid and cystine are autoclaved together in the same solution the ascorbic acid will be oxidized and destroyed. Both glucose and ascorbic acid are liable to be destroyed if heated in the presence of alkali.

Purity of Materials

Needless to say, only materials of the highest purity should be used in the preparation of defined media. It is necessary, however, to realize that even the purest chemicals have traces of impurities which may be important. Heavy metals are particularly toxic to cells and thus care must be taken with materials which may have been purified by lead or mercury precipitation. On the other hand, very small traces of certain substances may be essential to the cells and probably the supply of some metals necessary for cells survival (e.g. zinc, manganese and cobalt) comes from minute amounts present in other salts. Some of the rare components of many media are only obtainable in purities of about 85 per cent, and the impurities may not be the same from batch to batch. Thus, in assessing how highly defined a medium is, it is essential to keep this matter of purity in mind.

Chemical instability

Several of the components added to synthetic media are rather unstable and the only way to store them effectively for long periods of time is in the dry state, preferably in the cold. A particular example of such a substance is glutamine which has an important place in cell metabolism. In solution, glutamine can be stored for a considerable time if it is kept frozen. This is true of most labile compounds so that it is important to prepare media fresh or if they are stored it should be at very low temperatures.

Stock solutions

From the above considerations, it is apparent that if large amounts of media are to made up for use and stored for any length of time difficulties are likely to be encountered and only extensive deep freezing facilities would permit the long-term storage of large amounts of medium. This procedure is often undesirable and the difficulty is usually overcome by making up media as a series of partially complete stock separately in whatever conditions are necessary for the individual components. Incompatible substances can likewise be kept separate until they are mixed together to make the medium.

Balanced Salt Solutions

All media used in tissue culture have a basic of a synthetic mixture of inorganic salts known as a '*physiological*' or *balanced salt solutions* (BSS). The function of this salt solution is to maintain the pH and osmotic pressure in the medium and also to provide an adequate concentration of essential inorganic ions. All physiological salt solutions have been derived from the salt solution originally described by Ringer. The first one to be developed specifically for supporting the metabolism of mammalian cells was Tyrode solution, and many other balanced salt solutions have been produced since. Most of them were devised with the object of preventing calcium deposition in the culture or in order to obtain better buffering conditions in the medium. In practice, none of these seems to have nay demonstrable advantage over any other from the point of view of the maintenance of cell growth. Some of them, however, do have advantages in ease of general handling. For instance, one disadvantage of the original Tyrode solution is that its preparation requires considerable care in order to avoid precipitation of calcium carbonate. The same criticism applies to one or two other salt solutions.

Table 9.1. Kreb's Balanced Saline Solutions

	Medium I	*Medium II*	*Medium III*
1. 09% NaCl (0.154 M)	80 parts	83 parts	95 parts
2. 1.15% KCl (0.154 M)	4 parts	4 parts	4 parts
3. 0.11 M $CaCl_2$	3 parts		3 parts
4. 2.11% KH_2PO_4 (0.154 M)	1 part	1 part	1 part
5. 3.82% MgSO4.H_2O	1 part	1 part	1 part
6. 1.3% $NaHCO_3$ (0.154M); trated with CO_2 until PH is 7.4 in medium I	21 parts	3 parts	3 parts
7. Na-phosphate buffer (100 parts of 0.1 M NA_2HPO_4 (1-78%Na_2 HPO_4.H_2O) and 25 parts of 0.1 M NaH_2PO_4 (1.38% NaH_2PO_4.H_2O)	18 parts		3 parts
8. 0.16 M Na-pyruvate (or L-lactase in medium I and II)	4 parts	4 parts	4 parts
9. 0.1 M Na-fumarate	7 parts	7 parts	7 parts
10. 0.10 M Na L-glutamate	4 parts	4 parts	4 parts
11. 0.3 M (5.4%) glucose	5 parts	5 parts	5 parts

Salt solutions can be divided into two types, those intended to equilibrate with air and those which are designed to equilibrate with a gas phase containing a high carbon dioxide tension of the order of 5 per cent. In the author's opinion the most useful type of the former is Hanks' balanced salt solution and the best example of the latter is Earle's balanced salt solution. Both of these have the great advantage that they are very easy to prepare by autoclaving of different parts. The best buffered salt solutions is undoubtedly Earle's solution but the fact that it requires a special gas mixture is sometimes a disadvantage.

The salt solutions developed by Krebs are also shown since these evolved empirically to give the best results in metabolic studies although they have not been employed to any extent in tissue culture work. Their most interesting feature is the inclusion of Krebs cycle intermediates

Materials

In making up salt solutions, it is advisable to use double glass distilled water. Ordinary tap water frequently contains a large number of ions many of which are quite harmless but some of which such as lead, are toxic to the cells. In addition, it may contain endotoxins released by the growth of gram negative bacteria, and double distillation from a glass vessel is desirable to remove these materials. It is a common practice to store glass distilled water in polythene bottles to prevent the reaccumulation of metallic salts from glass, as has been shown to occur by Healy. Of course, many metallic ions are also introduced as impurities of the salts used in the solutions. The salts should, therefore, be of the highest analytical purity and special bottles should be kept aside for making up balanced salt solution since, particularly in a chemical laboratory, it is possible for them to be contaminated by other materials weighed simultaneously. Phenol red is added to many salt solutions as a pH indicator since it has been found to be non-toxic in concentrations up to 0.005 per cent.

Preparing a Balanced Salt Solution

The main difficulty encountered in preparing a balanced salt solution is the formation of a precipitate of calcium and magnesium carbonate and phosphate. Certain salt solutions are more prone to do this than others and the authors prefers Earle's and Hanks' solutions since they give little trouble of this kind. The avoidance of precipitation constitutes the on only special art involved in preparing a salt solution.

1st Method

The calcium and magnesium salts are weighed and dissolved in about 100 ml water. All other components are weighed and dissolved in about 800 ml water. While this large amount of solution is stirred vigorously, the calcium and magnesium salt solution is slowly added. The volume is adjusted to 1 litre and the solution filtered.

2nd Method

The salt solution is made up as before but sodium bicarbonate is omitted. 1.4 percent sodium bicarbonate is prepared separately. These two solutions are sterilized by autoclaving. After cooling the volume of the main solution is adjusted with distilled water if necessary and bicarbonate solution added to complete it. Note that when sodium bicarbonate is autoclaved, carbon dioxide is evolved and sodium carbonate remains. If the bottle is sealed tightly before autoclaving the carbon dioxide is, of course, retained. Otherwise, sodium bicarbonate can be reconstituted if desired by bubbling sterile carbon through the solution after it has been cooled.

Use of stock solution

It is convenient to prepare stock solutions of salt mixtures. A solution is made up with ten times the concentration of all components except sodium bicarbonate, which is omitted. This is stored in a polythene bottle. If it is subsequently to be autoclaved it can be preserved with a little chloroform (which will be drive off during sterilization). To reconstitute the BSS, distilled water is added to dilute the stock solution ten times. 1.4 per cent, sodium bicarbonate is used, as above to adjust the pH after sterilization.

PARTIALLY COMPLETE AND COMPLETE 'SYNTHETIC' MEDIA

The nutritional requirements of cultured cells, so far as they are known, have already been discussed, as have also the factors that have to be taken into account in making up a mixture. A few of the established media illustrated in the accompanying tables will now be considered before the manufacture of an actual synthetic medium is illustrated by a practical example.

Table 9.2. Fishcer's Supplementary Medium

	Milligrams per 1000 ml.		*Milligrams per 1000 ml.*
Glucose	800	Ribovlalvin	0.2
Mannose	100	Pyridoxine	0.3
Galactose	100	Choline	10.0
Inositol	20	Panthothenic acid	0.07
Adenosinetriphosphate	200	Biotin	0.007
Fructose diphosphate	100	*p*-Aminobenzoic acid	1.0
α-Glycerophosphate	100	Ascorbic acid	2.0
Inosinic acid	30	Methylnaphthahydroquinone	0.005
L-Lysine	15	L-Arginine	5
Nicotinic acid	0.3	L-Histidine	2
Cozymase	5.0	DL-Valine	14
Glutathione	5.0	L-Leucine	9
Creatine	10.0	DL-Isoleucine	10
Hypoxanthine	100.0	DL-Threonine	12
Sodium succinate	10.0	DL-Phenylalanine	7
Sodium fumarate	10.0	L-Tryptophan	2
Sodium malate	10.0	DL-Methionine	6
Sodium oxaloacetate	10.0	Cystine	5
Glutamine	250.0	Thiamine	3.0

Attempts to prepare a synthetic medium were made in the earliest years of tissue culture but the first systematic studies were initiated by Albert Fischer who, using dialyzed plasma as a basal medium, showed that the small molecular fraction could be adequately replaced by a mixture of amino-acids similar to those found in fibrin. His first papers were published in 1941 and culminated in the publication of his medium V-605 in 1948. White independently proposed a similar type of medium in 1946 and these media can be recognized as precursors of the defined media used today. Fisher's medium is shown here mainly for historical interest.

Table 9.3. White's Medium (1949) For Chick Heat Muscle

	Milligrams per 1000 ml.		*Milligrams per 1000 ml.*
Glucose	8,500	DL-Valine	130.0
NaCl	7,000	DL-Isoleucine	104.0
KCl	375	DL-Phenylalanine	50.0
$Ca(NO_3)_2\ H_2O$	210	Glycine	100.0
$MgSO_4$	275	Cystine HCl	1.0
$Na_2HPO_4 12H_2O$	145	Glutathione	1.0
KH_2PO_4	26.0	Thiamine	0.1
$NaHCO_3$	550.0	Pyridoxine HCl	0.5
$Fe(NO_3)_3\ 9H_2O$	1.3	Niacin	0.5
L-Lysine HCl	156.0	Roboflavin	0.1
L-Arginine HCl	78.0	Biotin	0.4
L-Histidine HCl	26.0	Inositol	0.5
L-Aspartic acid	60.0	Choline	1.0
L-Glutamic acid	140.0	Carotene	0.1
L-Leucine	156.0	Vitamin A	0.1
L-Tryptophan	40.0	Ascorbic acid	0.5
L-Proline	50.0	Calcium phantothenate	0.1
L-Cystine	15.0	Folic acid	0.05
DL-Threonine	130.0	β-Alanine	0.05
DL-Methionine	130.0	Phenol red	5.0

Following in the Fischer tradition Morgan, Morton and Parker produced medium 199 in 1950. This medium is still widely used, particularly in the maintenance of tissue for virus production in vaccine manufacture. As can be seen from Table 9.4, it is very complex, containing many substances in addition to those which have been proved to be essential. Despite its complexity this medium is unable to maintain cells for more than a day or two by itself. However, if serum is added it can maintain many cell strains indefinitely. Morgan has subsequently produced a modification of 199, known as M150. The essential differences are the use of Hanks' BSS instead of Earle's

BSS as a basis for the medium, replacement of xanthine by monosodium xanthine and of the barium salt of ATP by the sodium salt, addition of glucose, glutamine and bicarbonate immediately before sterilization by filtration and the addition of antibiotics.

Table 9.4. Morgan, Morton and Parker's Medium No 199 (1950)

	Milligrams per 1000 ml.		*Milligrams per 1000 ml.*
L-Arginine	70.0	Riboflavin	0.010
L-Histidine	20.0	Pyridoxine	0.025
L-Lysine	70.0	Pyridoxal	0.025
L-Tyrosine	40.0	Niacin	0.025
DL-Tryptophan	20.0	Niacinamide	0.025
DL-Phenylalanine	50.0	Pantothenate	0.01
L-Cystine	20.0	Biotin	0.01
DL-Methionine	30.0	Folic acid	0.50
DK-Serine	50.0	Choline	0.50
DL-Threonine	60.0	Inositol	0.05
DL-Leucine	120.0	p-Aminobenzoic acid	0.05
DL-Isoleucine	40.0	Vitamin A	0.10
DL-Valine	50.0	Calciferol (Vit.D)	0.10
DL-Glutamic acid	150.0	Menadione (Vit. K)	0.01
DL-Aspartic acid	60.0	α-Tocopherol phosphate	
DL-Alanine	50.0	(Vit. E)	0.01
L-Proline	40.0	Ascorbic acid	0.05
L-Hydroxyproline	10.0	Glutathione	0.05
Glycine	50.0	Cholesterol	0.2
Cysteine	0.1	Tween 80 (oleic acid)	20.0
Adenine	10.0	Sodium acetate	50.0
Guanine	0.3	L-Glutamine	100.0
Xanthine	0.3	Adenosine triphosphate	0.2
Thymine	0.3	Ferric nitrate	0.1
Uracil	0.3	Ribose	0.5
Thiamin	0.010	Deoxyribose	0.5

Following the introduction of medium 199 many media of increasing complexity have been produced and two of these, medium 858 and medium NCTC 109, are illustrated because they are both capable of supporting the growth of certain cell strains in the total absence of added serum or other biological fluids. According to Dr. Raymond Parker, the cell strains involved are not normal cell strains but have adapted themselves to grow in these media. The media are even more complex than medium 199 and there can be little doubt that they contain many materials that the cells do not require.

Table 9.5. Synthetic Medium No. 858

	Milligrams per 1000 ml.		*Milligrams per 1000 ml.*
Amino-Acids		TPP (88% pure)	1.0
L-Arginine	70.0	FAD (60% pure)	1.0
L-Histidine	20.0	UTP (90% pure)	1.0
L-Lysine	70.0	Glutathione (100% pure)	10.0
L-Tyrosine	40.0		
L-Tryptophan	20.0	*Lipid Sources*	
L-Phenylalanine	50.0	Tween 80 (oleic acid)	5.0
L-Cystine	20.0	Cholesterol	0.2
L-Methionine	30.0	*Nucleic Acid Derivatives*	
L-Serine	50.0	Adenine deoxyriboside	10.0
L-Threonine	60.0	Guanine deoxyriboside	10.0
L-Leucine	120.0	Cytosine deoxyriboside	10.0
L-Isoleucine	40.0	5-Methylcytidine	0.1
L-Valine	50.0	Thymidine	10.0
L-Glutamic acid	150.0	*Miscellaneous*	
L-Aspartic acid	60.0	Sodium Acetate	50.00
L-Alanine	50.0	D-Glucuronic acid	3.6
L-Proline	40.0	L-Glutamine	100.0
L-Hydroxyproline	10.0	D-Glucose	1,000.0
Glycine	50.0	Phenol red	
L-Cysteine	260.0	(pH indicator)	20.0
Vitamins		*Ethanol*	
Pyridoxine	0.025	(as an initial solvent for fat-soluble	
Pyridoxal	0.025	constituents)	16.0
Biotin	0.01		
Folic Acid	0.01	*Antibiotics*	
Choline	0.50	Sodium penicillin G (added just	
Inositol	0.05	before use)	1.0
p-aminobenzoic acid	0.05	Dihydrostreptomycin sulphate	100.0
Vitamin A	0.10	n-Butyl parahydroxybenzoate	0.2
Ascorbic acid (Vit. C)	50.0	*Inorganic Salts*	
Calciferol (Vit. D)	0.10	NaCl	6,800.0
α-Tocopherol phosphate		KCl	400.0
(Vit.E)	0.01	$CaCl_2$	200.0
Menadione (Vit. K)	0.01	$MgSO_4.7H_2O$	200.0
Coenzymes		$NaH_2PO_4.H_2O$	140.0
DPN (95% pure)	7.0	$NaHCO_3$	2,200.0
TPN (80% pure)	1.0	Fe, as Fe $(NO_3)_3$	0.1
CoA (75% pure)	2.5		

Table 9.6. Protein-free Chemically Defined Medium NCTC 109

	Milligram per 1000 ml	*Approx. Equiv. in millimoles*
L-Alanine	31.48	0.35
L-Alpha amino butyric acid	5.51	0.05
L-Arginine	25.76	0.15
L-Asparagine	8.09	0.06
L-Aspartic acid	9.91	0.07
L-Cystine	10.49	0.04
L-Cysteine	26.00	0.21
D-Glucosamine	3.20	0.02
L-Glutamic acid	8.26	0.05
L-Glutamine	135.73	0.93
Glycine	13.51	0.08
L-Histidine	19.73	0.13
Hydroxy-L-proline	4.09	0.03
L-Isoleucine	18.04	0.14
L-Leucine	20.44	0.16
L-Lysine	30.75	0.21
L-Methionine	4.44	0.03
L-Ornithine	7.38	0.06
L-Phenylalanine	16.53	0.10
L-Proline	6.13	0.05
L-Serine	10.75	0.10
L-Taurine	4.18	0.03
L-Threonine	18.93	0.16
L-Tryptophan	17.50	0.09
L-Tyrosine	16.44	0.09
L-Valine	25.00	0.21
Thiamine hydrochloride (Vitamin B_1)	0.025	
Riboflavin	0.025	
Pyridoxine hydrochloride (Vitamin B_6)	0.0625	
Pyridoxal hydrochloride	0.0625	
Niacin	0.0625	
Niacinamide (Nicotinamide)	0.0625	
Pantothenate, calcium salt dextrorotatory	0.025	
Biotin	0.025	
Folic acid	0.025	
Choline chloride	1.25	
i-Inositol	0.125	
p-Aminobenzoic acid	0.125	
Vitamin B_{12}	1.00	
Vitamin A (Crystalline alcohol)	0.25	

Calciferol (Vitamin D)	0.25
Menadione (Vitamin K)	0.025
Alpha tocopherol phosphate, disodium salt (Vitamin E)	0.025
Glutathione—Monosodium salt	10.10
Ascorbic acid	49.90
Cysteine hydrochloride	259.90
Diphosphopyridine nucleotide (Coenzymase, coenzyme 1)	7.0
Triphosphopyridine nucleotide—Sodium salt	1.0
Coenzyme A	2.5
Cocarboxylase	1.0
Flavin adenine dinucleotide	1.0
Uridine triphosphate—Sodium salt	1.0
Deoxyadenosine	10.0
Deoxycytidine—HCl	10.0
Deoxyguanosine	10.0
Thymidine	10.0
5-methyl-cytosine	0.1
Tween 80	1.25
Glucuronolactone	1.8
Sodium glucuronate	1.8
Sodium acetate	50.0
Phenol red	20.0
Sodium chloride	6800.0
Potassium chloride	400.0
Calcium chloride	200.0
Magnesium sulphate	200.0
Sodium monobasic phosphate	140.0
Sodium bicarbonate	2200.0
Glucose	1000.0

Many attempt have been made to simply them and to identify the essential components. The most accurately defined medium of this sort is undoubtedly the one devised by Eagle. However, it is unable to support growth in the total absence of added biological fluid and before considering it Waymouth's medium MB 752/1 may be examined. It is, in a way, a compromise. It is much simpler than most of the other synthetic media but it contains rather more components than Eagle's medium since this contains only those which are essential for minimal growth rather than those for optimal growth. Previously Dr. Waymouth devised some simple media which would support cell-growth with the addition of pure albumin and peptone but without the inclusion of serum.

Medium MB 752/1, in the absence of any addition, has supported the rapid growth of a subline of the strain L mouse subcutaneous fibroblast through many passages.

Table 9.7. Waymouth's Medium MB 752/1

	Milligram per 1000 ml	*Equivalent in millimoles*
NaCl	6000	103
KCl	150	2.0
$CaCl_2.2H_2O$	120	0.82
$MgCl_2.6H_2O$	240	1.18
$MgSO_4.7H_2O$	200	0.81
Na_2HPO_4	300	2.11
KH_2PO_4	80	0.59
$NaHCO_3$	2240	26.7
Glucose	5000	27.8
Ascorbic acid	17.5	0.1
Choline. HCl	250	1.8
Cysteine. HCl	90	0.57
Glutathione	15	0.05
Hypoxanthine	25	0.18
Glutamine	350	2.38
The above ingredients, made up at double the stated concentrations, in glass distilled water, constitute solution BNI. Prepared by dissolving all components except the three buffer salts *first*, in about 80 percent, of the water. Then add the buffer salts in a small volume of water. This prevents precipitation of Ca or Mg phosphates.		
Thiamine HCl	10	0.03
Ca Pantothenate	1.0	0.003
Riboflavin	1.0	0.003
Pyridoxin, HCl	1.0	0.003
Folic acid	0.4	0.0008
Biotin	0.02	0.00008
m-Inositol. $2H_2O$	1.0	0.005
Nicotinamide	1.0	0.008
Vitamin B_{12}	0.2	0.00015
A stock solution of the B vitamins is conveniently made at 40 times the above concentrations.		
L-Cystine	15	0.06
Glycine	50	0.66

L-Phenylalanine	50	0.30
L-Glutamic acid	150	0.02
L-Aspartic acid	60	0.46
L-Tyrosine	40	0.22
L-Lysine. HCl	240	1.42
L-Proline	50	0.44
L-Methionine	50	0.34
L-Threonine	75	0.64
L-Valine	65	0.55
L-Isoleucine	25	0.19
L-Leucine	50	0.38
L-Tryptophan	40	0.20
L-Arginine. HCl	75	0.36
L-Histidine. HCl	150	0.80
NaOH	to pH 7.4	2.5

The amino acids are conveniently made up in a stock at ten times at above concentrations.

Medium 752/1 is made by adding to 37.5 ml water, 50.0 ml of BNI+2.5 ml of the stock B vitamin solution + 10.0 ml of the stock amino acid solution (total 100 ml).

Eagle medium is the simplest of the defined media in general use. It has been established that all its components are essential for cell growth and that most of the other components of more complicated media are not. Being relatively simple it provides a useful example of method of preparing a '*synthetic*' medium and this will be described in detail.

Preparation of Eagle's Medium

As has been indicated earlier, all complicated media are prepared as a series of stock solutions, these being combined immediately before the medium is required for use. The reasons for grouping materials together will be described in parenthesis in each case.

Stock Solution 1. Hanks' balanced salt solution, which glucose and bicarbonate omitted. (By leaving out these two substances it is possible to sterilize the salt solution by autoclaving. Also, glucose can combine with some amino-acids and alkaline conditions can induce degradation of various substances so that it is undesirable to have either present in making up some of the other stock solutions.)

Stock Solution 2 (×100). Dissolve the following L-amino acids in 100 ml. of stock solution 1 by heating to about 80°C:0.174 g. arginine;

0.032 g. histidine; 0.182 g. lysine; 0.131 g. leucine; 0.262 g. isoleucine; 0.075 g. methionine;; 0.083 g. phenylalanine; 0.119 threonine; 0.020 g. tryptophan; 0.117 g. valine. (This is general amino-acid stock solution, containing the more soluble and more stable amino-acids).

Table 9.8. Eagle's Medium (Slightly modified).

	Milligrams per 1000 ml.	*Approx Equivalent in millimoles*
L-Arginine	17.4	0.1
L-Cystine	6.0	0.05
L-Histidine	3.2	0.02
L-Isoleucine	26.2	0.2
L-Leucine	13.1	0.1
L-Lysine	18.2	0.1
L-Methionine	7.5	0.05
L-Phenylalamine	8.3	0.05
L-Threonine	11.9	0.1
L-Tryptophan	2.0	0.01
L-Tyrosine	18.0	0.1
L-Valine	11.7	0.1
L-Glutamine	146.0	1.0
Choline	1.0	
Nicotinic acid	1.0	
Pantothenic acid	1.0	
Pyridoxal	1.0	
Riboflavine	0.1	
Thiamine	1.0	
i-Inositol	1.0	
Biotin	1.0	
Folic acid	1.0	
Glucose	2000.0	
NaCl	8000.0	
KCl	400.0	
$CaCl_2$	140.0	
$MgSO_4.7H_2O$	100.0	
$MgCl_2.6H_2O$	100.0	
$Na_2HPO_4.2H_2O$	60.0	
KH_2PO_4	60.0	
$NaHCO_3$	350.0	
Phenol red	20.0	
Penicillin	0.50	

Stock Solution 3 (×100). Dissolve the following L-amino acids in 0.1 N HCl: 0.18 g. tyrosine; 0.06 g. cystine. (These two amino-acids are less soluble than the others and are most conveniently made up separately in dilute acid).

Stock Solution 4 (×100). Dissolve the following B vitamins in 100 ml. of stock solution 1 : 0.1 g. choline; 0.1 g. nicotinic acid; 0.1 g. panthothenic acid; 0.1 g. pyridoxal 0.01 g. riboflavin; 0.1 g. thiamine (aneurin); 0.1 g. (This is a general stock solution of the more soluble B vitamins. Since they are required in very small amounts, it is more convenient to prepare a strong stock than to weigh out the very small amounts required for a more dilute stock solution).

Stock Solution 5 (×100). Dissolve the following B vitamins in 100 ml. of stock solution 1 by the addition of a new drops of 0.5 N NaOH (until the pH is about neutral as judged by the colour of the phenol red): 0.01 g. biotin; 0.01 g. folic acid. (These two vitamins are much less soluble than the rest and therefore cannot be made up at 1,000 times working strength with the other B vitamins).

Stock Solution 6 (×100). Dissolve 10 g. glucose in stock solution 1. (Glucose is best kept out till the end since it can form complexes with some of the amino acids if heated together with them. Also if it is kept as a separate stock solution it is easy to adjust the glucose concentration in the medium as desired).

Stock Solution 7. Dissolve 1.4 g. sodium bicarbonate in 100 ml. distilled water. (The majority of labile compounds are most stable in slightly acid conditions than in slightly alkaline conditions. Thus it is best to keep alkali out of the medium until the end).

Stock Solution 8 (×100). Dissolve the following antibiotics in 100 ml. distilled water : 200,000 units sodium penicillin-G; 0.5 g streptomycin sulphate. (These being labile, are best made up separately and stored in small aliquots frozen solid).

Stock Solution 9 (×100). Dissolve 1.46 g. glutamine in 100 ml. stock solution 1. (Glutamine is also labile and is best made up in the same way and stored in small aliquots frozen solid).

Stock solution 1-7 are quite stable and they can safely be stored at ordinary refrigerator temperatures. It is advisable to sterilize them by filtration through sintered glass before doing so. As has already been mentioned, the remaining solutions are best stored in the frozen condition after sterilization. (It is as well to note that labile chemicals, such as glutamine, should be stored in cool, dry conditions). The easiest way to do this is keep them in sealed jars, containing silica gel, in the refrigerator or deep-freeze cabinet). Each week a 'working stock solution' is made up from the above stock solutions by combining them in the following quantities:

Stock solution 2-10 ml.; stock solution 3-10 ml.; stock solution 4-1 ml.; stock solution 5-10 ml.; stock solution 6-10 ml.; stock solution 8-10 ml.; stock solution 9-10 ml.

This provides 61 ml. of 'working stock solution', which is enough to make up 1 litre of medium. It can be stored for a week or so in the refrigerator without loss of potency.

To make up 100 ml. of growth medium containing, say, 10 per cent, of horse serum, the following are mixed immediately before use; 84 ml. stock solution 1;6.1 ml. 'working stock solution'; 10 ml. horse serum. Stock solution 7 (sodium bicarbonate) is then added to adjust the pH to between 7.2 and 7.4. While the complete medium can be stored for a few days in the refrigerator this practice is not to be recommended since the more labile compounds begin to undergo degradation almost immediately.

The above version of Eagle's medium is the one used at Glasgow and is taken from several papers published by Dr. Eagle and his colleagues. When fresh serum is being used inositol can safely be omitted. However, the medium as described will support adequate growth for some time even when supplemented with exhaustively dialyzed serum and it is therefore useful for many experimental procedures.

The principles employed in making up Eagle's medium are identical with those used for the more complicated media.

Other Synthetic Media

Many media is addition to those mentioned have been described in the literature. Most of them are not generally used but two others are of special interest and will therefore be discussed.

The first of these is shown in Table 9.9 and was evolved by G.A. Fischer for the cultivation of the L-5178 Y ascites tumour of mice. It deserves mention because, although it is far from rigorously defined, it is the only medium currently available which seems to support the continuous propagation of this strain. The L-5178 Y cell is particularly interesting since it is of lymphoblastic type and grows spontaneously in suspension. In the author's experience this medium has proved a very useful general nutrient solution for the cultivation of exacting cell strains.

Table 9.9. Constituents of Culture Medium for L-5178 Lymphoblasts

	Medium concentration
	(gm/l.)
(a) NaCl	8.0
KCl	0.4
$MgCl_2.6H_2O$	0.1
Na_2HPO_4	0.06
$NaH_2PO_4.H_2O$	0.067
(b) Glucose	1.0
(c) $NaHCO_3$	1.0
(d) 5% acid hydrolyzed casein + tryptophan 0.5 mg/ml.	5 ml./l.
	mg/l.)
(e) Glycine	20
Cystine	7.5
Histidine	20
(f) Glutamine	200
(g) Vitamins:	
Thiamin. HCl	1.0
Nicottinamide	0.5
Ca-Pantothenate	0.5
Pyridoxal, HCl	0.5
D-Ribose	0.5
Riboflavin	0.5
Choline chloride	1.5
i-Inositol	1.5
(h) Biotin	0.01
(i) Ascoric acid	1.6
Glutathione (reduced)	1.5
(j) Serum	2-10%
(k) Peptone	0.06%
(l) Folic acid	10 mg/l.
(m) Penicillin	100 units/ml.
Stereptomycin	0.050 mg/ml.
(n) Phenol red	10 mg/l.
(o) $CaCl_2.H_2O$	0.182 gm/l.

The other medium of special interest is that devised for adult mammalian organ cultures by Trowell and designated medium T8. This protein-free medium is extremely simple in composition and has been used to maintain fragment of many adult tissues for from some days to a few weeks in a healthy condition. It must be borne in mind that the function of this medium is quite different from that of media for rapidly-growing cell strains because the organ fragments have a relatively low mitotic rate (sometimes no mitoses at all) and they are maintained *in vitro* for a relatively short period of time).

Table 9.10. Trowell's Medium T8 (1959)

	Milligrams per 1000 ml.	*Approx. equiv. in millimoles*
NaCl	6100	104
KCl	450	6
CaCl	220	2
$MgSO_4.7H_2O$	250	1
$NaH_2PO4.2H_2O$	450	3
$NaHCO_3$	2820	33.5
Glucose	4000	22
L-Arginine HCl	21	0.1
L-Cysteine HCl	47	0.3
L-Histidine HCl	10	0.05
L-Isoleucine	26	0.2
L-Leucine	26	0.2
L-Sysine HCl	36	0.2
L-Methionine	7.5	0.05
L-Phenylalanine	16.5	0.1
L-Threonine	24	0.2
L-Tryptophan	4	0.02
L-Tyrosine	18	0.1
L-Valine	23	0.2
Thiamine HCl	17	0.05
P-Aminobenzoic acid	35	0.25
Insulin	50	0.001
Chloramphenicol	30	0.1
Phenol red	10	0.03

Media for the cultivation of tissues from cold-blooded animals, insects and plants

By far the greatest volume of work in tissue culture has been done with tissues from warm blooded animals. However, tissues from a great many other sources have been successfully grown and there seems no reason to believe that any animal or plant tissue will prove completely refractory to attempts to grow *in vitro*.

Naturally, the more divergent the species the more divergent are the nutritional requirements likely to be and when we examine entirely different phyla we find considerable differences in the basic needs for growth of tissue in vitro. However, having drawn attention to this obvious fact, it must be remarked that the most extraordinary thing about known nutritional requirements of different cells is their general similarity. Even between plants and animals the differences in requirements for inorganic ions are quantitative rather than quantitative and the greatest divergencies are in the requirements for vitamins and amino-acids. Even these reflect greater degrees of metabolic autonomy rather than fundamentally different needs. Some of the more complex chemical molecules such as certain of the hormones which are characteristic for different organisms do not seen to be necessary for the survival of unorganized (and even, in some cases, organized) cell growth.

Culture of Tissue from Cold-Blooded Animals

The cells of many cold-blooded animals, particularly of embryonic and larval stages, will survive for some time at 18°C in a suitable salt solution, due to the fact that they often contain stores of food materials. Salt solutions for cold-blooded animals are similar to those for mammalian and avian tissues, differences being due mainly to different osmotic pressure requirements.

Table 9.11. Balanced Salt solutions for Cold-blooded Animals (Grams per litre)

Substance	*Holtfreter (Amphibia and fish)*	*Frog Ringer (Amphibia)*	*Modified Locke (Insects)*	*Carlson (Grass-hoppers)*
NaCl	3.50	6.50	9.00	7.00
KCl	0.05	0.14	0.42	0.20
$CaCl_2$	0.10	0.12	0.25	0.02
$MgCl_2$				0.10
$MgSO_4$		0.20		
NaH_2PO_4				0.20
$NaHCO_3$	0.20		0.20	0.05
Glucose			2.50	0.80

In order to support prolonged survival of cells from these animals natural media, analogous to those used for warm-blooded animals, have been employed. Both fish and amphibian tissue have been grown in media consisting of the appropriate salt solution supplemented with serum or plasma and embryo extract. In both cases avian serum and embryo extract have been used successfully but some workers have considered homologous media preferable. In this respect it is worth remembering that the first effective tissue cultures were made from tissues from a cold-blooded animals and the medium considered of clotted frog lymph.

Blood may be obtained from cold-blooded animals by the same procedures as for warm-blooded animals. Cardiac puncture is used for extracting blood from fish and certain large shell-fish such as the lobster. Cannulation is also employed, particularly for amphibia. The aorta of the salamander may be cannulated similarly.

Instead of chick embryo extract, extracts of tadpoles have been used for amphibian material and extract of fish try for fish tissues. It is not unlikely, however, that synthetic media of the type used for mammalian cells will prove more suitable than these.

Insect Tissue Culture

Similar considerations apply to the cultivation of insect tissues and cells. Again, the basis of the medium is always a salt solution providing the inorganic ions in a mixture which produces a suitable osmotic pressure and pH. As is the case with most other kinds of cell culture, it is necessary to supplement the balanced salt solution with some material of biological origin in order to obtain prolonged growth. In the case of insect tissues the material used for this purpose is haemolymph. This is obtained by cutting off the last two or three segments from a large pupa and expressing the fluid. All haemolymph tends to darken when it is exposed to air and this is associated with the development of toxic properties. In order to inhibit tyrosinase and thus prevent this darkening, a number of expedients have been employed. Commonly the haemolymph is saturated with phenylthiourea. This method is successful for many purposes, but Wyatt has found that phenylthiourea is inhibitory to the growth of cells and finds that better results are obtained when tyrosinase is precipitated from haemolymph by heat treatment. This consists of raising the haemolymph to 60° for five minutes, deep freezing for 24 hours and then centrifuging at 6,000 g. for ten minutes, the supernatant being used.

Mainly due to the pioneer work of Wyatt, synthetic supplements have been developed for use with insect materials. Wyatt's medium, which is shown in Table 9.12, was developed on the basis of an analysis of the haemolymph of the silkworm and with it she obtained good growth of silkworm cells. This medium has now been improved by Grace by the addition of some vitamins of the B group. When Grace's medium, illustrated in Table 9.13, is supplemented with about 3 per cent, of heat-treated haemolymph (concentrations up to 50 per cent,

Table 9.12. Wyatt's Medium for Insect Tissues

	Milligrams Per 1000 ml.	*mM*
1. *Inorganic Salts*		
NaH_2PO_4	1,100	8
$MgCl_2.6H_2O$	3,040	15
$MgSO_4.7H_2O$	3,700	15
KCl	2,980	40
$CaCl_2$	810	7.2
2. *Sugars*		
Glucose	700	3.9
Fructose	400	2.2
Sucrose	400	1.1
3. *Organic Acids*		
Malic	670	5
α-Ketoglutaric	370	2.5
Succinic	60	0.5
Fumaric	55	0.5
4. Amino-Acids		
L-Arginine HCl	700	3.3
DL-Lysine HCl	1,250	6.9
L-Histidine	2,500	15.7
L-Aspartic acid	350	2.63
L-Asparagine	350	.65
L-Glutamic acid	600	4.08
L-Glutamine	600	4.11
Glycine	650	8.66
DL-Serine	1,100	10.5
DL-Alanine	450	5.05
β-Alanine	200	2.25
L-Proline	350	3.2
L-Tyrosine	50	0.27
DL-Threonine	350	2.94
DL-Methionine	100	0.67
L-Phenylanine	150	0.9
DL-Valine	200	0.7
DL-Isoleucine	100	0.77
DL-Leucine	150	1.44
L-Trypotophan	100	0.61
L-Cystine	25	0.1
Cysteine HCl	80	0.5

have also been used) excellent growth of silkworm cell is obtained, and Dr. Grace has succeeded in keeping some insect cells alive in it for about a year.

Table 9.13. Grace's Medium for Insect Tissues

	Milligrams per 1000 ml.	*mM*
1. *Inorganic Salts*		
NaH_2PO_4	1,100	8
$MgCl_2.6H_2O$	3,040	15
$MgSO_4.7H_2O$	3,700	15
KCl	2,980	40
$CaCl_2$	810	7.2
2. *Sugars*		
Glucose	700	3.9
Fructose	400	2.2
Sucrose	400	1.1
3. *Vitamins*		
Thiamine	0.01	
Riboflavine	0.01	
Ca pantothenate	0.01	
Pyridoxine	0.01	
Nicotinic acid	0.01	
Biotin	0.01	
Folic acid	0.01	
Choline	0.01	
m-inositol	2.00	
4. *Amino-Acids*		
L-Arginine HCl	700	3.3
DL-Lysine HCl	1,250	6.9
L-Histidine	2,500	15.7
L-Aspartic acid	350	2.63
L-Asparagine	350	2.65
L-Glutamic acid	600	4.08
L-Glutamine	600	4.11
Glycine	650	8.66
DL-Serine	1,100	10.5
DL-Alanine	450	5.05
β-Alanine	200	2.25
L-Proline	350	3.2
L-Tyrosine	50	0.27
DL-Threonine	350	2.94
DL-Methionine	100	0.67
L-Phenylanine	150	0.9

DL-Valine	200	0.7
DL-Isoleucine	100	0.77
DL-Leucine	150	1.44
L-Trypotophan	100	0.61
L-Cystine	25	0.1
Cysteine HCl	80	0.5
5. *Miscellaneous*		
Malic acid	670	5
α-Ketoglutaric acid	370	2.5
Succinic acid	60	0.5
Fumaric acid	55	0.5
Cholesterol	30	

Plant Tissue Culture

The nutrition of plant cells was examined in detail and described accurately some years ago, so that completely synthetic media for the growth of plant tissues have been available for a long time. The reason for this is mainly that the problem of plant cell nutrition is considerably simpler than the problem of animal cell nutrition. Plants have a very much more versatile metabolism than animals and can use simpler molecules for almost all their synthetic processes. Thus, in addition to an inorganic salt mixture (corresponding to the xylem sap) the only materials required for good growth are sucrose, a mixture of vitamins and some glycine (corresponding to the essential components of the phloem sap). White's nutrient solution for plants is of this very simple form.

Table 9.14. White's Medium for Plant Tissues

	Mg./litre
$Ca(NO_3)_2$	200
$MgSO_4$	360
Na_2SO_4	200
KNO_3	80
KCl	65
NaH_2PO_4	16.5
KI	0.75
$Fe_2(SO_4)_3$	2.5
$MnSO_4$	4.5
$ZnSO_4$	1.5
H_3BO_3	1.5
Glycine	3.0
Thiamine	0.1
Niacin	0.5
Pyridoxine	0.1
Sucrose	20,000

A more complicated solution has been devised by Gautheret and his co-workers. This contains many of the trace elements and cysteine instead of glycine. White's medium is adequate for roots and many plant tumour tissues as well as a number of other materials. On the other hand, Gautheret's solution is widely used for normal tissues. As a rule no addition to these media is required, but in the case of a few plant tissues a natural supplement to the synthetic mixtures has been found useful. Supplements used have been yeast extracts and coconut milk.

Table 9.15. Gautheret's Solution

	Mg./litre
$Ca(NO_3)_2$	100
KNO_3	25
$MgSO_4$	25
KH_2PO_4	25
$Fe_2(SO_4)_3$	50
$MnSO_4$	2
KI	0.5
$ZnSO_4$	0.1
H_3BO_3	0.15
$Ti_2(SO_4)_3$	0.2
$NiSO_4$	0.05
$CoCl_2$	0.05
$CuSO_4$	0.05
Glucose	30,000
Agar	6,000
Cysteine–HCl	10
Thiamine	1
Ca panthothenate	0.1
Biotin	0.1
Inositol	100
Naphthalene acetic acid	0.3

10

PREVENTION OF CONTAMINATION

Media for the cultivation of cells and tissues are highly nutritious not only for animal cells but also for bacteria and fungi. The majority of these microorganisms have a much more rapid growth-rate than cells and frequently produce toxins which are lethal to them. Hence, the most important part of tissue culture technique comprises the avoidance of combination and the growth of tissue in aseptic conditions.

Contamination of materials can be prevented in two ways:

1. *Sterilization*. This term implies the removal of microorganisms already present.
2. *Aseptic Technique*. This term implies the prevention of contamination of materials already sterile.

In tissue culture work the sources of contamination are as follows:

1. The apparatus.
2. The culture medium.
3. The tissue itself.
4. The atmosphere in which the operation is performed.
5. The operator.

Apparatus is rendered bacteria free before commencing work by sterilization. Likewise the medium is usually sterilized before use but some components are prepared aseptically. The tissue is usually obtained in a sterile condition by the application of aseptic principles but occasionally it is necessary to sterilize it. Contamination from other sources is prevented almost entirely by the application of the aseptic technique.

STERILIZATION PROCEDURES

Sterilization can be achieved by :

1. Physical destruction of microorganisms, i.e., by dry heat, moist heat and radiation.
2. Chemical destruction of microorganisms, i.e. by antiseptics and antibiotics.`
3. Physical removal of microorganisms, i.e. by filtration, centrifugation and washing.

Most *apparatus* is sterilized by dry heat or by moist heat. In general, apparatus which is not damaged by high temperatures is sterilized by dry heat since this is most convenient method. Where excessive heat can damage the material, most heat is used. It is more efficient than dry heat because of the height latent heat of steam which rapidly transfers its energy to the materials being sterilized. Autoclaving (sterilizing by steam under pressure) is frequently used for solutions, rubber ware, cloth and so on. Simple boiling is used for instruments and is sufficiently reliable for the vast majority of procedures. Sometimes it is impossible to use heat sterilization, for example, in the case of certain plastics which soften at temperatures about 100°C. In these cases antiseptics are used but they should be of a relatively innocuous type and readily volatile. Seventy per cent, ethyl alcohol is the one most commonly used. Ultraviolet radiation can also be employed for some pieces of apparatus, e.g., plastic rays.

Solutions are nearly all amenable to filtration but some, such as plasma and embryo extract, cannot be satisfactorily filtered and these have to be prepared aseptically. A few simple solutions which are not damaged by heat can be autoclaved. Since proteins are coagulated at high temperatures, this procedure cannot be applied to serum, embryo extract, plasma, or other media of biological type. However, salt solutions can be autoclaved provided certain precautions are taken. These have been described in the section on synthetic media.

Solutions can also be sterilization by ultraviolet radiation. The requires special apparatus and has not been widely applied in tissue culture. Antibiotics are frequently added to growth media but they should not be added with the object of killing organisms present, rather should they be added as an auxiliary safety measure in case of accidental contamination.

For *laboratory benches* and *tables*, some sort of antiseptic is used. Again, alcohol is commonest and benches can be swabbed down with it to reduce the number of bacteria present. The most valuable measure in sterilizing surfaces of this sort is the laying of dust which may otherwise be disturbed during working and settle on the apparatus. For this reasons, many people prefer to wipe benches before use with a thin oil which forms a slightly adhesive layer and prevents dust being disturbed. It is a good practice to have laboratory floors treated with oil every few days for this reason. Bench tops can also be sterilized by ultraviolet radiation but in this case it should always be remembered that only those surfaces on which the radiation falls directly will be sterilized. All areas within shadows remain unaffected.

The sterilization of *air* can be achieved in a number of ways. As a general rule, it is not necessary to sterilize air since contamination from this source can be prevented by ordinary aseptic technique. However as discussed in the next chapter, if the project in hand necessitates a source of sterile air, this can be achieved most satisfactorily by a filtration system, conveniently combined with an airconditioning until. Air can also be sterilized can also be sterilized with ultraviolet radiation by arranging an ultraviolet lamp so that all the air in the room is sterilized for a period before it is used. If this is done then precautions must be taken to ensure that the lamp is not switched on when the room is being used since eye injuries may result. Air can also be sterilized by precipitating dust and microorganisms by means of an aerosol. In order to produce an aerosol, it used to be the practice in many laboratories to fill the aseptic room with a cloud of steam which was allowed to settle before commencing work. This creates conditions of humidity within the room and this is undesirable since it favours the growth of fungi and the survival of mites. Another type of aerosol that has been used in ethylene glycol which is vapourized by placing beaker containing it on a hot-plate or in a hot bath.

Fig. 10.1. Ultraviolet lamp of a type convenient for air sterilization.

Sterilization by Dry Heat

Most apparatus and almost all glassware can be sterilized by this method. An oven is used and it is desirable, if possible, to use one with a forced air circulation. However, any oven will suffice and a domestic kitchen oven is quite suitable. It is important to raise all the contents of the oven to a sterilizing temperature for a suitable period of time and the material to be sterilized should be arranged in such a way that air can circulate freely.

The disadvantage of this type of sterilization is that dry air being a poor conductor of heat relatively cool pockets may remain within the contents of the oven. Consequently, a rather high temperature for a rather long time is recommended and 90 minutes at 160°C is usually employed. It is sometimes useful to mark the glassware with a temperature-sensitive paint to show that this temperature has, in fact, been reached. Several kinds of such paints are available (e.g. Thermocolor by Griffin & George).

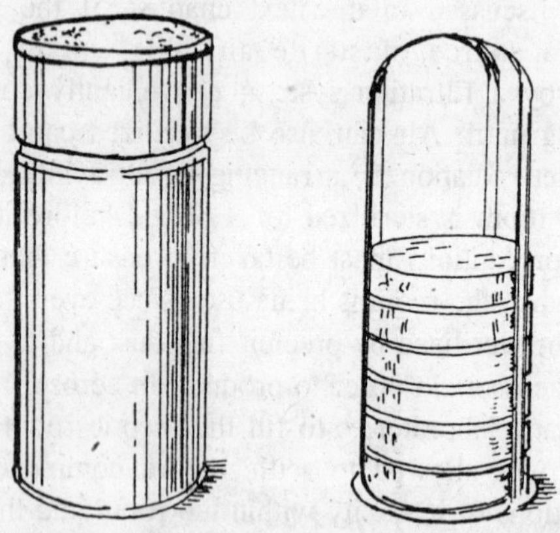

Fig. 10.2. A useful type of can for sterilization of Petri dishes.

Materials put into the oven are conveniently sterilized in tin containers, e.g. biscuits tins. However, before using tins for sterilizing glassware they should be placed open in the oven and heated to sterilization temperature for a period of two or three hours. This drives off various volatile materials which would otherwise be transferred to the glassware. It should be noted that some tins are lacquered inside and these are not suitable since the lacquer is burned or volatilized and the products are deposited on the glassware.

Fig. 10.3. Glass pipette holders of this type are particularly useful for Pasteur pipettes and can be made to order.

Glass articles can be wrapped in aluminium foil or in paper for sterilization. Pure craft paper should be used and not the artificial variety. If it is overheated craft paper becomes brittle and tends to release volatile tars which are transferred to the glass. Some care is therefore necessary to avoid this. A special paper is available in the United States and is superior to craft paper for heat sterilization.

Sterilization by Moist Heat

Sterilization by moist heat is usually carried out in the autoclave. A domestic pressure cooker is sometimes more convenient for sterilizing small amounts. It operates on exactly the same principle as the autoclave. The old vertical autoclave is not so satisfactory as the modern horizontal type which should be used, if possible, Sterilization can be carried out in autoclave at 5,10 or 15 lbs, pressure and the recommended procedure is to autoclave at 15 lbs. Pressure for 20 minutes. Solutions should be sterilized for rather shorter time, about 15 minutes as a rule. In older type of autoclaves it is necessary to control sterilization by time and pressure, but it is better to control it by temperature and in most modern autoclaves this can be done by reading the thermometer situation in the steam outlet line. A temperature of 115°C should be maintained for about 15 minutes. Almost all known organisms are destroyed within about one minute in such conditions but it is necessary to maintain them for 15-20 minutes because air pockets may exist within the sterilization packages or steam may be unable to reach various parts for other reasons. In case inadequate steam penetration occurs, it is advisable to use sterilization controls of some sort. These are readily available from any surgical supplier. They are placed in the package and change of colour indicates adequate sterilization.

In operating the autoclave, it is essential to ensure that all air is displaced by steam before the pressure is allowed to rise. In modern autoclaves this occurs naturally because of the design. In older autoclaves it is essential, after closing the lid, to allow steam to escape for a considerable time. If it is allowed to escape under water it is possible to tell when most of the air has been removed, since the steam produced is immediately dissolved and this gives rise to a loud, rattling noise. When this point has been reached the autoclave is closed and the pressure allowed to rise to 15 lbs. After 20 minutes the gas or electricity is shut off and the pressure and temperature allowed to fall over another 20 minutes to half an hour. By thus gently lowering the pressure, solutions are prevented from boiling. In using the autoclave, the same considerations have to be kept in mind as in sterilizing by dry heat. In particular the packages should be so arranged that steam will gain access to all parts.

Materials may be prepared for sterilization in the autoclave by wrapping either in gauze or in paper. For this purpose, again craft paper can be used, but much better is Patapar, which has already been mentioned in connection with sterilization by dry heat. This material is tough and does not release any noxious materials. In the case of large glass vessels, if the entire vessel is not enclosed in this way, all orifices should be closed by means of cotton wool plugs or by paper or aluminium foil coverings..

Solutions should be sterilized in bottles or flasks, stoppered with plugs of cotton wool wrapped in gauze. Alternatively, they can be fitted with rubber stoppers or screw-caps but these should be left loose during autoclaving and tightened afterwards. Cotton wool plugs are preferable since, as the solution cools, air is sucked into the vessels and this is filtered by the cotton. With the other method there is always at a risk, although quite a small one, of sucking organisms into the medium.

Moist heat in the form of boiling water can be also used for sterilization. Very few organisms will survive boiling water for more than a few seconds and therefore this technique is very convenient for rapid sterilization of syringes and instruments. They should be protected from bumping by wrapping them in gauze.

Radiations

Radiations can be also be used for sterilization. Ultraviolet light has been used, for instance, in the sterilization of plastic trays used for metabolic inhibition tests. It has to be used with care, however, since shadowed areas are unaffected. In the future it is possible that gamma rays and neutrons may be used for sterilization of materials. However, it should be remembered that these radiations are capable of producing chemical changes in the substances irradiated. At the present moment they have rather limited uses and the only general application of radiation is in the ultraviolet sterilization of cubicles and rooms, particularly where pathogens such as viruses are handled.

Antiseptics

As has already been mentioned above, certain *antiseptics* are quite commonly used for swabbing down laboratory benches and shelves and for this purpose 70 per cent alcohol is the safest since not only is it volatile but it is relatively non-toxic, at least in low concentrations. Ether can also be used as an antiseptic in this way, but it is not recommended since it is rather volatile and inflammable. The only other chemical substance commonly used for sterilization is chloroform which is frequently added to solutions to preserve them, e.g. stock solutions of balanced salt solution. The chloroform is driven off during autoclaving.

Plastic dishes can also be sterilized with alcohol. The dishes are washed with 95 per cent, alcohol, then placed in paper bags while wet and left in an incubator to dry, when they are ready for use.

Antibiotics

Very many *antibiotics* have been used in tissue culture work and most of the available data is summarized in the table. It is bad practice to become completely dependent on antibiotics but at the same time, nothing is to be gained by trying to conduct routine tissue culture without them. It should be observed, however, that many antibiotics are quite toxic to cells and in some cases the levels of toxicity approach the effective levels.

By far the most useful general antibiotic is *penicillin*. It is usually added as sodium penicillin G to media to give a final concentration of 20 to 50 units per ml. At these concentrations it is completely harmless to all cell types and is inhibitory to the vast majority of bacteria. As a routine the author recommends that only penicillin should be used in media but if contamination by penicillin-resistant organisms is to be expected then *streptomycin* may also be used and should be added to give a final concentration of 50 μg. per mol. Fungal infections may be inhibited by the use of *Mycostatin* (Nystatin). This fungistatic agent should be added to give a final concentration of 20 μg per mol. It should be noted that mycostatin is unstable and has usually deteriorated almost completely after 24 hours in tissue culture medium at 37°C. In the author's opinion it is better to retain these latter two antibiotics for use in an emergency if a contamination should arise in a valuable strain, since sometimes such contaminations can be eliminated by the use of high concentrations of antibiotics.

In addition to their routine use in media, antibiotics can be used at much higher concentrations to sterilize contaminated tissues prior to explanation.

Filtrations

Many different kinds of filters are available. They can be divided into four main groups:

1. Asbestos pads.
2. Kieselguhr and porcelain filters.

3. Sintered glass filters (fritted glass filters).
4. Millipore filters.

Asbestos pads of the Seitz type have the great merit that they effectively remove all microorganisms but permit a fairly high rate of filtration. The disadvantage is that in some cases at least they release toxic substances which are subsequently inhibitory to growing cells. Most filters, especially washed filters, seem to be satisfactory, but some kinds are not suitable for filtering media for tissue culture, unless very large quantities are to be treated, in which case it becomes worthwhile to subject the filter to very thorough washing. Many types of apparatus are available. Seitz filtration is particularly suitable for sterilization of serum.

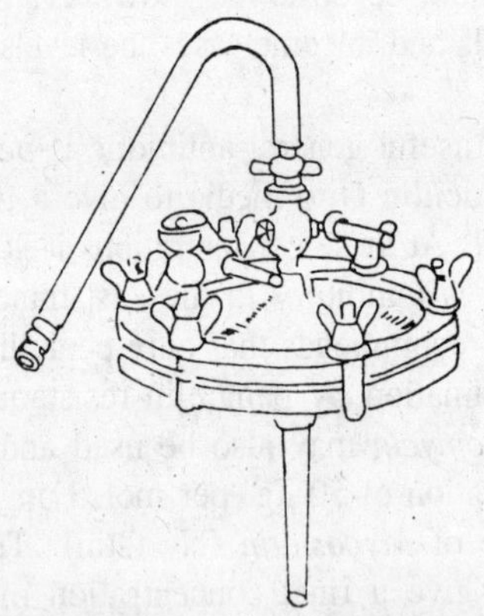

Fig. 10.4. Apparatus for filtration with asbestos pads by pressure or vacuum.

Kieselguhr and *porcelain* filters are probably the best filters for general purposes. In this group the Selas filter is particularly useful. It can be treated in exactly the same way as the Berkefeld filter but is much more uniform and has a higher rate of filtration. In both cases a grade should be used which is guaranteed to remove all microorganisms. New filters should be prepared by scrubbing lightly with a soft brush and then flushing thoroughly in both directions with several litres of ordinary tap water followed by flushing with 200-300 cc. of deionised water. They can be sterilized by autoclaving. After use they should be thoroughly washed again in both directions with a very large amount of water. If any protein remains this can be removed by a crude trypsin or pancreatin solution.

When filters become really dirty, they can be cleaned by incinerating in a muffle furnace. The temperature should be raised

Fig. 10.5. Apparatus for pressure filtration with filter candles.

slowly to 650°C over a period of three or four and allowed to drop back to normal overnight. It is essential that the filters should be thoroughly dry before being put in the muffle furnace, otherwise they will be destroyed. All new filters and all filters which have been cleaned by this technique of incineration should be tested. This is done by submerging each filters in water and introducing air under pressure into it. At the pressure use for filtration no air should normally escape. At a higher pressure a uniform bubbling of air should arise from all over the filter. The appearance of large bubbles from any area indicates defects. For some filters, e.g. the Selas filter, a bubbling pressure is given by the manufacturers.

Sintered glass filters are useful for small quantities of material. They have the disadvantage of filtering slowly. Again bacteriological grades should be used. A No. 5 filter supported on a No. 3 filter is most suitable and this should be referred to as 3/5 filter. The United States equivalent in fritted glass filters in the UF grade.

Millipore filters can be used in place of Seitz filters. They have the advantages of containing no impurities and providing rapid rates of filtration. Also the used filters can be discarded and thus cleaning procedures need not be considered. They are rather expensive at present.

In filtering biological materials, it is desirable to use pressure rather than suction. When suction is employed, it causes carbon dioxide to be evolved from the solutions and this gives rise to a radical change in pH. Also protein solutions forth very badly when a vacuum is applied. Filtration by pressure avoids both difficulties.

The other physical means of removing bacteria, centrifugation and washing, are relatively inefficient but are not to be despised on occasions. Heavily contaminated fluids such as bacterial digests, can be clarified by centrifugation before filtering. Thorough washing of materials with distilled water or sterile BSS will remove micro-organisms very effectively unless they are strongly absorbed to or embedded in the material. Contaminated cultures can sometimes be cleared in this way and cultures should always be washed before trying to eliminate contamination by adding antibiotics to the medium

Storage of Sterile Materials

It is convenient to sterilize materials in large batches and then to store them until they are required. Two difficulties may arise, contamination during storage and deterioration due to absorption of toxic substances from the atmosphere.

For these reasons *solutions* are best kept in tightly sealed containers rather than containers stoppered with cotton wool since otherwise they may readily absorb substances such as formalin or ammonia. The most convenient and cheapest containers for this purpose are medical prescription bottles ("medical flats") which can be obtained in a great variety of sizes, with silicone liners if desired, and already sterilized (although it may not always be desirable to rely on this). Alkaline solutions should never be stored for a long time in glass, however, since the glass is slowly dissolved and heavy metal ions appear in solution.

Clean sterile *glassware* is subject to the same two dangers. It is not uncommon for laboratories to be lightly infested with mites carried by laboratory animals and these can gain access to stored glassware through paper wrapping and cotton wool plugs. This is sometimes the cause unexplained heavy mixed contaminations. Also, particularly in chemical laboratories, stored glassware gradually acquires a deposit of chemicals on the surface with the result that cells will no longer grow in them. Thus it is advisable not to store glassware for more than a week or two and it should be carefully protected from possible contamination by careful wrapping. The best system to employ is sterilization of glassware and apparatus in envelopes of nylon film. This film is supplied in the form of a tube from ¼ in. to 20 ins. wide and in rolls of suitable lengths. The material to be sterilized is placed in a section of this and the ends are folded over and sealed with a special autoclave adhesive tape to form a closed bag. The packages are then sterilized by autoclaving in the usual manner. They nylon film is permeable to steam but not to bacteria so that after removal from the autoclave the contents remain sterile until required. These materials are obtainable from Portland Plastics Ltd., Bassett House Hythe, Kent. For sealing packages for autoclaving Scotch hospital autoclave tap no. 222 is particularly useful since it is printed with a heat sensitive ink which gives positive evidence that sterilization has been adequate.

11

Cell Culture in Plants

When a plant is wounded, cells at the wound site proliferate into an undifferentiated mass called *callus tissue*. This tissue can be thought of as a kind of scar tissue and, in nature, callus formation seals the wound against attack by pathogenic bacteria and fungi. Callus is probably most familiar to you as the swollen, knot-like growth seen on trees at the site of limb removal, but herbaceous plants as well as woody plants produce callus. Plant cell culture exploits this natural wound response of plants by encouraging callus growth on small pieces of excised tissue or "explants".

On artificial media, callus can be maintained in a state of persistent, undifferentiated growth. By changing the concentration of plant hormones in the medium, however, the callus of many plant species can be made to redifferentiate into whole plants. The ability of single cells to regenerate whole plants is referred to as *totipotency*.

The commercial applications of plant tissue culture techniques are quite important. Regeneration of plants from cell culture offers a practical strategy for plant cloning, since all regenerants from a culture should be genetically identical. For commercially valuable plants that are difficult, costly or inefficient to propagate by cuttings or other asexual means, cell culture sometimes offers the only practical means of propagation. Most hybrid orchids, for example, are propagated today by the tissue culture method of "meristemming" or "mericloning". Plant cultures are also being investigated as sources of valuable plant products like drugs, flavors and fragrances.

Many strategies for genetic engineering of plants rely on plant tissue culture. Plant cells in culture can be genetically transformed by a number of techniques. Explants can be transformed directly by *Agrobacterium tumefaciens* or by bombardment with DNA-coated particles from a particle gun. Protoplasts (plant cells with their cell walls removed) are the targets of microinjection and electroporation (DNA uptake mediated by an electric field). Plant cell culture is central to these techniques in that cell culture allows transformants to proliferate and, sometimes, regenerate into genetically identical clones.

While many formulations for plant tissue culture media have been developed, all contain the same basic types of ingredients. These are:

1. Inorganic salts
   Macro and micronutrients
2. Vitamins
3. An organic carbon source

In addition to these components, most formulations contain plant growth regulators (plant hormones), *auxin* and *cytokinin*. Callus media are generally solidified by the addition of agar. (Plant cultures are sometimes maintained as a suspension of cells in liquid medium, but this type of culture will not be included in this laboratory experiment.)

The medium used in this experiment, MS medium, is a formulation developed by Murashige and Skoog (1962.) MS medium is one of the most commonly used of all plant culture media. The ingredients included in this medium are:

1. Inorganic Salts	mg/l
Macronutrients	
Ammonium nitrate	1650
Potassium nitrate	1900
Calcium chloride (anhydrous)	332
Magnesium sulfate (anhydrous)	180.7
Potassium phosphate	170
Micronutrients	
Ethylenediaminetetraacetic acid (EDTA, Disodium Salt)	37.3
Ferrous Sulfate (heptahydrate)	27.8
Manganese sulfate	16.9
Zinc sulfate (heptahydrate)	8.6
Boric acid	6.2
Potassium iodide	0.83

Sodium molybdate (dihydrate)	0.25
Cobalt chloride (hexahydrate)	0.025
Cupric sulfate (pentahydrate)	0.025
2. Vitamins	
Myo-inositol	100.00
Thiamine hydrochloride	0.40
3. Carbon source	
Sucrose	30,000.00

The medium is solidified with 7.5 g/I of agar. The amounts of plant growth regulators added to the medium are variable, depending on whether the culture is to be maintained as callus or made to regenerate whole plants. The medium used in this experiment contains Indole Acetic Acid (IAA, an auxin) at a concentration of 1.0 mg/I. No cytokinins are added. The effects of varying the concentrations of plant growth regulators on the cell culture will be investigated in another module.

Safety Guidelines

Follow standard laboratory safety practices.

1. Preparation of media requires an autoclave or pressure cooker. Use of this equipment requires care to avoid serious burns.
2. An alcohol or gas flame is used to sterilize instruments during the lab exercise. Due care should be used with open flames. Precautions should be taken to avoid igniting hair or clothing.

Experimental Outline

1. Disinfect plant tissue and prepare explants.
2. Inoculate medium with plant tissue.
3. Observe cultures over a period of several weeks.

Materials

Sterile media (this has been prepared prior to class)
Plant material
10% Household bleach solution (prepare fresh)
Sterile Water
70% Ethanol
95% Ethanol
Forceps
Scalpel

Pre-lab Preparation

Tissue culture medium should be prepared in advance of the laboratory period. This can be done several days to a week ahead of time. Prepared media can be stored at room temperature. Plant materials used for the experiment should be purchased or collected shortly before the class meeting.

Preparation of plant material for culture, i.e. surface sterilization, will take approximately 30 minutes. Plan another 30 minutes for novices to prepare explants and inoculate the culture medium. The time required for students to complete the exercise will depend on whether students are waiting for space in a laminar flow hood.

Depending on the length of the laboratory period, you may want to combine this exercise with Module 3: Regeneration and Cloning. Both exercises require that the students initiate cell cultures.

Because results from both Modules 1 and 3 require several weeks of observation after the initial laboratory class period, these exercises are best used early in the semester.

Timetable of events

Students will learn how to establish plant cell cultures from a variety of plant tissues. In accomplishing this, students learn how to prepare plant tissue for culture, practice sterile technique, and learn about the components of plant culture media.

The "bare bones" methods described in the student guide for this module (and for others in the plant tissue culture unit) are adaptable to most lab conditions. Depending on the availability of sterile transfer hoods, specialty glassware, etc., you may want to elaborate on the protocols presented, but very little specialized equipment is actually required for success.

Equipping the work area

Plant tissue culture is normally done in a laminar flow hood, a containment hood, or in a transfer cabinet. The laminar flow hood has a box shape with one open side facing the worker. Fans in the hood circulate air through a particle (HEPA) filter which removes airborne contaminants. Air flow in the laminar flow hood is from the back of the box outward past the worker. Sterility in the work area is maintained by this flow of clean air, but can be compromised by introducing contaminated equipment into the hood, by leaning into the hood while working, or by placing non-sterile items between open cultures and the back of the hood.

Containment hoods, like laminar flow hoods, depend on air flow to maintain sterility in the work area. Because these hoods are designed to prevent the escape of cultured cells from the work area, however, the opening at the front of the hood is partially closed (generally by a window) and air travels from the top of the cabinet downward to intakes on the work surface. Air does not flow out of the hood into the lab environment without first being filtered to remove contaminants. The protection offered to workers by a containment hood is unnecessary for most plant tissue culture purposes, but the hood is very satisfactory for handling plant cultures.

Transfer cabinets (or closely-related glove boxes) are box-like enclosures, sealed on the top and on three sides, with limited access from the front. Sterility is maintained within the cabinet only to the degree that air movement from outside the box to the inside is restricted. The transfer cabinet is really no more than a shelter against airborne contaminants. When used properly, the transfer cabinet is an economical and effective enclosure for cell culture. If desired, a transfer cabinet can be economically "homemade" from plywood or plastic. Painting the plywood surface with enamel will make it washable.

In the absence of a sterile hood or transfer enclosure, plant culture experiments can still be carried out. However, depending on the cleanliness of the lab environment and the sterile technique of the students, losses to fungi and bacteria can be expected. The best results will be obtained if windows are kept closed to restrict air movement and if bench tops are wiped with 70% ethanol before work begins. Perhaps most important of all, students should wash their hands well before doing any culture work.

The equipment required for plant cell culture is modest. For handling cultures, the work station should be outfitted with a Bunsen burner, a covered container of 95% ethanol, forceps (long enough to reach into culture flasks or tubes), and a sharp scalpel. For disinfecting the work area, a wash bottle or spray bottle of 70% ethanol and a supply of clean (preferably autoclaved) wipers or paper towels will suffice. For media preparation, you will need a balance weighing to 0.1 mg. a pH meter, and an autoclave. Plant cell cultures can be grown in petri dishes, baby food jars, test tubes, Erlenmeyer flasks, mason jars, etc. The only requirements for a suitable container are sterility (they should be autoclavable or presterilized/disposable) and they must be able to be opened and closed repeatedly without introducing contaminants. Cotton or foam stoppers can be used as closures. Specially-designed caps for tubes, flasks and baby food jars are also available. Ideally, plant callus cultures are incubated at 26-28°C, but they will grow at slightly cooler room temperature. Callus cultures can be maintained either in the light or in the dark. In the classroom situation, a drawer in the lab bench can serve the purpose of incubator. Callus cultures from which plants are being regenerated should be maintained under fluorescent lights with a day length of 16 hours.

Preparation

Obtain plant material: A number of different plants can be tested in this experiment. Students may even want to try some materials of their own. The best material for explants comes from plants raised indoors or in the greenhouse because they seem to be "cleaner". Some vegetable tissues are notable exceptions to this. Some plants to try are: Tobacco stem or leaf, carrot roots, African violet leaf or petiole, cauliflower florets. If you want to use vegetables from the grocery, try to purchase these as fresh as possible. Plants other than those listed above will also do very well. Make the selection of explants part of the experiment.

Make media: The recipe for MS medium is written out in the student guide to the laboratory. If you choose to assemble the medium from stock bottles on the lab shelf, it will be easier to prepare a 100 x concentrated stock solution of ingredients listed as micronutrients and keep this stock in the refrigerator. Also, plant growth regulators are more easily handled as stocks. Dissolve IAA in a small amount (few mls) of 1 M NaOH, then dilute to 1.0 mg/ml with distilled water. Dissolve kinetin in a small volume of 1 M HCl, then dilute to 1.0 mg/ml with distilled water. Just before autoclaving, adjust the pH of the plant medium to 5.8.

Pre-prepared media are available commercially and offer a simple and cost-effective alternative for media preparation. For this experiment, order "MS salts with Minimal Organics". To this powder, you will only have to add sucrose (table sugar from the grocery store will work; add 30 g/l), agar (7.5 g/l), and plant growth regulators (For this experiment, add only IAA; 1 mg/l).

Method

1. Prepare plant material for culture. Select fresh-looking, healthy (not brown, bruised or wilted) plant material. This can be leaf, stem, or root. Cut the tissue into manageable pieces. These should

be small enough to fit into a beaker for disinfecting, but large enough so that they can be trimmed after disinfection.

2. Move your work into a sterile transfer hood. If a hood is not available and you are working at the lab bench, wipe the work area with a solution of 70% ethanol before beginning and try to keep your work covered as much as possible. Before proceeding to disinfect your plant tissues, wash your hands thoroughly.
3. Wash the explants in a beaker of distilled water to which you have added a few drops of detergent.
4. Using forceps, transfer the explants to a 70% solution of ethanol for 2 minutes.
5. Again using forceps, transfer the explants to a 10% solution of household bleach for 5-10 minutes. Tender leaf tissue should not be left in the bleach for more than about 5 minutes. Root or stem pieces can be left for longer times.
6. After treatment with bleach, the tissue is considered to be sterile. Care should be taken to avoid re-contaminating it. At your work station should be a covered container of 95% ethanol and a Bunsen burner. Before they are used to handle sterile plant tissues, forceps and scalpels should be dipped in ethanol and passed through the flame several times to sterilize them. Be careful as you do this. An alcohol flame is nearly colorless and therefore invisible. Take precautions to avoid igniting hair or clothing. If you should accidentally set the container of ethanol on fire with the hot instruments, extinguish the flame by replacing the cover on the container.

   Note: Instruments are sterilized by ethanol and not by heat or flame.
7. Rinse the explants in three, 5 minute changes of sterile distilled water. Keep the tissue in its last rinse until you are ready to use it.
8. Transfer the explant from the rinse water to the lid of a sterile petri dish. For leaf tissue, cut the tissue into small squares, not larger than 1 cm^2 using a flamed scalpel. Cut edges that were in direct contact with ethanol and bleach should be trimmed away. Do not use leaf tissue that appears "soaked through" by the disinfectant. Root or stem tissue should be trimmed to remove any tissue that was in direct contact with the disinfecting solutions. Cut the tissue into small cubes, 0.5-1.0 cm^3. In preparing your

explants for culture, be mindful of the need to maintain their sterility. Flame your forceps and scalpel frequently and don't lay them down on the work surface after flaming. When working in a laminar flow hood, remember that it is important not to place any contaminated materials up wind of your work. If you are not working in a hood, work quickly and keep your work covered as much as possible.

9. Using flamed forceps, transfer the explants to the culture medium. Be careful to prevent contaminating the medium during this procedure. Explants should be placed firmly in contact with the medium, but should not be buried in it.
10. Label your cultures as to source of explant (type of plant and tissue), date, and your name. Consult the instructor about where cultures are to be incubated.
11. Check the cultures periodically before your next lab class. Any cultures that become grossly contaminated should be autoclaved before disposal.

Results

Examine your cultures and others. Use a microscope. Describe the callus tissue formed and determine the following. How do callus cells compare in appearance to the cells from which they grew? Is there any difference in the appearance of the cultures developed from different plant species or from different tissues of the same plant? Did all of the cultures grow equally well? Did callus tissue form at the same time and grow at the same rate in each of the cultures? Did any of the cultures begin to regenerate roots or shoots? What types of contaminants appeared in the cultures?

Within the first week explants can be expected to swell noticeably. Callus formation should begin during this time. Callus should appear on cut edges of the explant and will look like a rough white fringe of cells. In subsequent weeks, the callus will proliferate into an undifferentiated mass of cells. If desired, the callus can be cultured away from the explant. Remove the clump from the explant tissue with sterile forceps and place it on fresh medium. Whether or not the callus is to be maintained, students should explore the culture with forceps to get a feel for the texture of this tissue. Depending on what plant material is used for the cultures, shoots or roots may begin to regenerate from the callus. This is not a likely outcome, but it is possible.

The major difficulty faced in initiating plant tissue cultures is that of contamination. Plant media are excellent substrates for a host of fungi and bacteria. Generally speaking, bacterial contaminants are more easily dealt with. Antibiotics such as cefotaxime (100 μg/ml) or piperacillin (100 μg/ml) can be included in media for short term bacterial control. If you choose to use antibiotics, remember that they should not be added to media before autoclaving. They should be added as a filter-sterilized stock solution to cooled medium immediately before it is dispensed into tissue culture vessels. Fungal contaminants are nearly impossible to get rid of. For the teaching lab, probably the best advice about contamination is to tolerate it if possible (that is, if the cultures aren't grossly contaminated) over the short term of the experiment and dispose of contaminated cultures by autoclaving. Sometimes, contaminants will not interfere with the outcome of the lab exercise.

12

Cell Line

Primary cell cultures are those that are composed of cells taken directly from a living animal. Primary cells normally contain a diploid set of chromosomes, have limited lifespan and undergo aging. In order to prepare a primary culture, an organ or tissue is removed from a freshly sacrificed animal and aseptically cut into small pieces and treated with enzymes such as trypsin, collagenase, pronase or combination thereof. After the cells have been separated from the tissue, they are inoculated into appropriate cell culture medium in tissue culture flasks. The cells will adhere to the surface of the flask and replicate until they come in contact with each other. Thus they attach and grow as a uniform layer of cells, or a monolayer, which is always one cell thick. These cells will stop growing once the surface is filled up and contact follows. This is known as contact inhibition. Cells can be cultured as stationary monolayers usually inoculated with 2×10^5 cells in 5 ml of cell culture medium in a petri dish designed for tissue culture work.

Some cell types can be grown in suspended state in culture medium. Suspended cells are generally derived from blood cells. B lymphocytes can be established in culture directly from leukemic cells or by infecting normal lymphocytes with Epstein-Barr virus. Suspension cell cultures will yield 5-10 times more cells per ml of medium than monolayer cultures.

It is fairly easy to derive a primary culture of fast growing cells like fibroblasts, and as a result most primary cultures consist of fibroblasts. Fibroblasts are connective tissue cells which secrete the extracellular matrix of connective tissue. In this exercise, primary fibroblast cultures will be derived from chicken embryo. They will be made by dissociating the entire embryo with the proteolytic enzyme trypsin. Embryo cells are somewhat easier to adapt to cell culture than adult cells, since embryo cells are less differentiated and more likely to be rapidly dividing. The culture that is first produced will contain a variety of cell types. It is possible, however, to eventually establish a homogeneous culture of fibroblasts, because the other cell types do not grow as fast and they will be diluted out in subsequent passages subcultures.

Safety Guidelines

Wear gloves, lab coat, and safety glasses. Standard laboratory safety procedures should be followed.

Experimental Outline

- Establish primary cell culture from chicken embryo.
- Culture cells as stationary monolayer in tissue culture petri dishes.
- Establish a homogeneous culture of fibroblasts.
- Grow cells in tissue culture medium under sterile conditions.
- Stain cells with trypan blue and count the number of unstained, live cells under a microscope using a hemocytometer.

Materials

Fertile chicken eggs, incubated 7-10 days.

A bottle of 0.25% trypsin solution.

A bottle of PBS (phosphate buffered saline).

Tissue culture flask 25 cm^2.

A bottle of HMEM (Hank's minimum essential medium) + supplemental 10% Fetal Bovine Serum (heat inactivated).

Three 100 mm plastic petri dishes.

One 50 ml plastic beaker.

150 ml squeeze bottle with 70% alcohol.

A sterile dissection kit containing sharp scissors, sharp scalpel and two pointed forceps.

95% ethanol.

5 ml plastic syringe.

Sterile gauze.

Two 125 ml flasks.

Trypsinizing flask.

15 ml centrifuge tubes (2).

Trypan blue (0.4%).

0.15 ml plastic microcentrifuge tube with an attached cap penicillin-streptomycin 100x stock solution.

1 ml and 5 ml pipets, sterile..

Pipet pumps.

CO_2 incubator.

Method

1. Keep 0.25% trypsin, HMEM and PBS in 37°C water bath.
2. Aseptically place 15 ml of PBS into each sterile petri dish.
3. Place the egg in a 50 ml beaker with its blunt end up, and disinfect the entire egg shell with 70% ethanol.
4. With sharp sterile forceps puncture the top of the shell and remove the shell all the way up to the base of the air cell.
5. Locate the position of the embryo and carefully tear the membrane away from the embryo with sterile forceps.

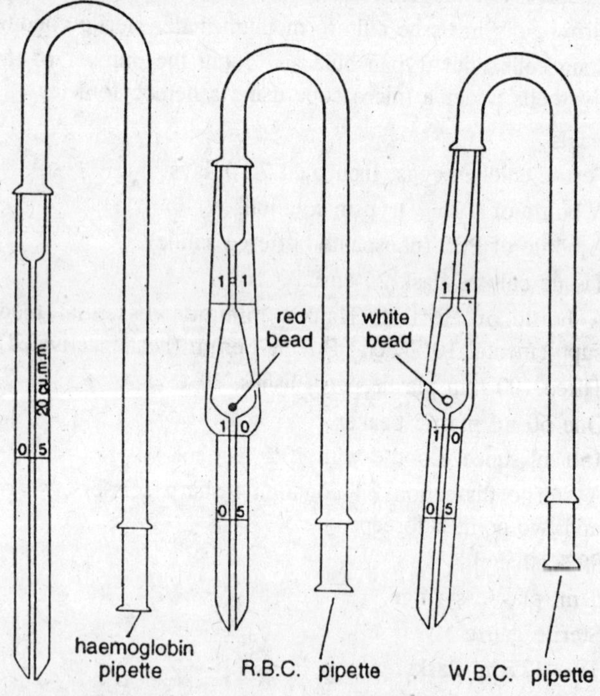

Fig. 12.1. R.B.C., W.B.C. and haemoglobin pipettes.

6. Insert a sterile pair of forceps into egg and grab embryo firmly. Take it out of egg and place in the first dish of PBS.
7. Cut off the head and feet of embryo with a sterile scissors. This procedure is done for 10 day old embryos but not for younger embryo.
8. With a flamed sterile forceps, transfer the remaining parts of the embryo to a petri dish containing PBS and rinse thoroughly. Repeat the rinse until no color is found in the rinse solution in the dish.
9. Take a sterile 5 ml syringe and a sterile capped test tube. Remove the plunger from the syringe and rest it on sterile kimwipes. Use a sterile forceps to transfer the remaining parts of the embryo from the petri dish into the syringe. Remove the cap from the tip of the syringe without touching the tip with your fingers. Uncap the sterile test tube and place the cap on sterile kimwipe and keep the tube in a test tube rack. Position the tip of the syringe over the test tube. After inserting the plunger back in the syringe, with a steady and even pressure, push the plunger all the way down so that the embryo is forced through the syringe into the test tube. The embryonic tissues are delicate and are homogenized to very fine pieces.
10. Transfer the tissue into a trypsinizing flask and add 10 ml of 0.25% prewarmed trypsin solution.
11. Place the trypsinizing flask on the shaker in the 37°C water bath and allow the suspension to swirl on the shaker at low speed so that the tissues of the embryo just barely graze the cutting edges of the trypsinizing flask. It will be necessary to continue this action for 15-60 minutes depending on the disaggregation process, and you will have to monitor this process closely. The flask is removed once the trypsin solution becomes cloudy with individual cells. Excessive trypsinization will reduce the cell viability significantly.
12. Allow fragments to settle; the supernatant is a cloudy suspension of cells. Pour this cell suspension through the side spout into a sterile 125 ml flask containing 10 ml of prewarmed HMEM with 10% Fetal Bovine Serum. The serum will inhibit further action of trypsin and prevent further damage to cells.
13. Obtain another sterile 125 ml flask and place two or thee layers of sterile gauze over the mouth of the flask and filter the cell suspension through this gauze to remove large cell clumps and other debris.

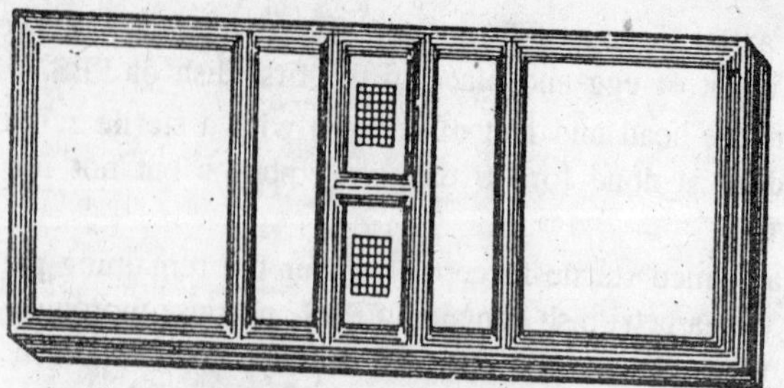

Fig. 12.2. The Burker Hawksley counting slide.

14. Take two clean sterile centrifuge tubes. Aseptically transfer the cell suspension into these tubes so that both tubes contain the same quantity. Pellet the cells by centrifugation in a table top centrifuge for 5 minutes at 1000 rpm. Discard the supernatant and resuspend the cell pellets in 5 ml of HMEM containing 10% Fetal Bovine Serum. Combine the cell suspension in one tube and cap it.
15. Remove the cap of the tube containing the cell suspension and by using a sterile 10 ml pipette, aspirate the suspension by drawing it up into the pipette and then forcing back into the tube. Do this about six times. It is important to keep the cell suspension uniform and the clumps broken up as much as possible. Determine the viability and cell density.
16. With a sterile 1 ml pipette, aseptically transfer 0.1 ml of cell suspension into 1.5 ml microcentrifuge tube.
17. Pipette out 0.9 ml of 0.4% (w/v) trypan blue solution into the above microcentrifuge tube (wear gloves when working with trypan blue). Cap the tube tightly and vortex it for 15 seconds. Immediately with a Pasteur pipette introduce a drop of it at the center of one edge of the cover glass covering one of the counting chambers of the hemocytometer. The fluid will be drawn under the cover glass by capillary action. Do not over fill the chamber.
18. Place the counting chamber on a regular microscope and focus with low power objective on the ruled area of the chamber. The hemocytometer is divided into 9 large blocks (1 mm^2), and each of these large blocks is divided into 16 smaller squares. Count the number of unstained cells (trypan blue penetrates damaged cells, and thus only stains those cells that are dead) in the central large block and four large corner blocks. If there are too many cells to count, make the necessary dilution.

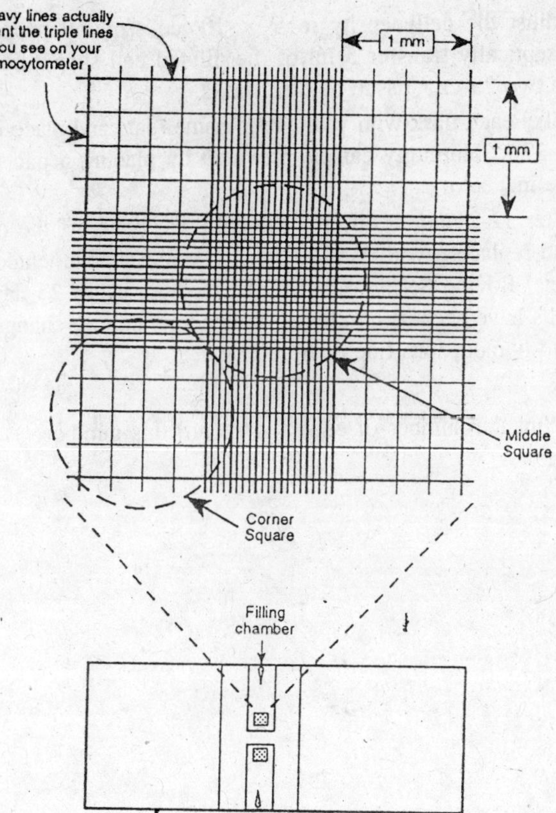

Fig. 12.3. Hemocytometer.

19. Divide total number of cells by 5 to obtain average number of cells per large block. (Note: if this number is smaller than 10 and larger than 100, the count will not be accurate.) Multiply by 10^4 and the dilution factor to obtain number of cells per ml in the suspension. An example is shown below:

Total cells counted in give large blocks	=	250
Average number of cells/block	=	50
Number of cells per ml	=	50×10^4

For the correct count, multiply the above number by 10 (dilution factor). The correct number of cells would be:

$$50 \times 10^5$$

$$5 \times 10^6$$

20. Adjust the cell density to 5×10^5 cells/ml by proper dilution. Aseptically transfer 5 ml of the diluted cell suspension into each of two 25 cm^2 tissue culture flasks.
21. Label each flask with your group name, date and place in incubator at 37°C Humidity can be provided by placing a pan of water in the incubator.
22. After 12-24 hours observe the cultures, pour off the old medium and replace it with 5 ml of fresh HMEM supplemented with 10% Fetal Bovine Serum. Change the medium every 23 days until the cells have become confluent. Notice how culture changes overtime into homogenous line of fibroblasts.

Results

Count the number of cells and record this number.

13

MAINTENANCE OF A CELL LINE

To culture and maintain established cell lines, a primary culture is split to produce new cultures and is subsequently known as a cell line. The two types of cell lines are (a) finite cell line and (b) continuous or established cell lines. A finite cell line is capable of a limited number of cell generations in vitro and after which cells die. A continuous cell line has the capacity for an infinite number of population doubling and thus in a sense immortal. Some continuous cell lines are also malignant, meaning they will grow as a tumor and invade other tissues if injected into an animal.

Earle is credited with establishing the first immortal cell lines in 1940. Most of these were established by treating rodent primary culture with chemical agents. One of these lines called L-cells was established in 1943 by treating primary mouse fibroblasts with the cancer causing agent methylcholanthrene. George and Margaret Gey derived cells from the cervical carcinoma of a black woman named Henrietta Lacks. This cell line which is called "HeLa", is the most common source of human cells used in research today. Today, many cell lines are available which can be obtained from American Type Culture Collection (ATCC), local medical colleges or any research institutions.

Established cell lines can be divided into two groups: those that can grow attached to a solid surface (monolayer) and those that grow in suspension. The mouse L-cells is a good example of the surface dependent cell line. When these cells are seeded in a culture vessel, they attach and grow to form a uniform layer of cells or a monolayer, which is always one cell thick. These cells will stop growing once they start filling up the surface and touching each other due to contact inhibition. In order to subculture a monolayer, it has to be dissociated to individual cells with trypsin, after which it can be put in a larger vessel with more space and fresh medium. Suspension cells are grown in liquid culture similar to bacterial culture.

Safety Guidelines

Wear gloves, lab coat, and safety glasses. Standard laboratory safety procedures should be followed.

Experimental Outline

- Maintain an established cell line in tissue culture
- Introduce Trypsin-EDTA into the culture flask, aseptically to help maintain conditions for a monolayer culture
- Observe individual cells using an inverted microscope
- Maintain cells, replace medium when a drop in the pH is observed based on change on phenol red indicator - changing from red to yellow

Materials

Chick embryo fibroblasts (from Module 11) or mouse L-cells

A bottle of PBS

A bottle of 1x Trypsin-EDT A solution (concentrated)

Trypan blue solution - 0.5% in PBS

A bottle of complete medium (HMEM), supplemented at 10% v/v with Fetal Bovine Serum and 1% v/v penicillin-streptomycin 100x stock

Waste container

10 ml pipets, sterile and pipet pump

CO_2 incubator

Method

1. Keep PBS, Trypsin-EDTA and complete medium in a 37°C water bath.
2. Observe the condition of the monolayer by holding the bottle up to the light. In a good culture the medium will appear clear and the cells will look smooth and confluent with no clumps
3. Observe the individual cells by using an inverted microscope at low power. Only those cells which undergo mitosis would be round and floating.
4. Make sure that the monolayer is confluent and if so, it is ready to be transferred.
5. Uncap culture flask and decant medium to a waste container.

6. Aseptically withdraw 10 ml of sterile PBS, and transfer it into the culture flask and recap the flask. Rock the flask gently from side to side for 10-15 seconds to rinse the cells. This step helps to remove the serum, which would inhibit the enzymatic activity of trypsin. Decant the solution into a waste container.
7. Add 0.2 to 0.5 ml of Trypsin-EDTA into the culture flask, swirl the trypsin around and incubate at 37°C for 3 minutes. Too much trypsin can damage cells. Cells are attached to cells and substrate by surface glycoproteins, serum proteins and chemical groups that coat the tissue culture-treated plastic surfaces of the flask. The enzyme trypsin catalyzes the hydrolysis of bonds involved in cell to cell and cell to substrate contact.
8. Once the cells begin to dislodge, tap the flask against the palm of your hand lightly. Hold the flask up to the light to check that no patches of monolayer remain. Observe the cells under an inverted microscope to confirm their presence.
9. Aseptically add 5 ml of complete medium. Proteins in the serum will stop the action of trypsin.
10. Aspirate the cell suspension by drawing it up into the pipette and expelling it back into the flask 5-6 times with the tip of the pipette pressed against the bottom corner of the flask. This will break up the cell clumps. Clumped cells will affect the accuracy of the cell count.
11. Count the cells using a hemacytometer.
12. Expand the culture by taking 1×10^5/ml for chicken fibroblasts or 1×10^4/ml for L-cells. If you are using 25 cm^2 flasks, transfer about 1 ml of cell suspension to each flask and add 9 ml of medium to the flask.
13. Incubate the culture at 37°C for 12-24 hours. After this, decant the old medium and add 10 ml of fresh medium and incubate. It is necessary to transfer (split) these cultures once they become confluent, which should occur in 3-4 days. Observe them under inverted microscope every 2-3 days. When the phenol red indicator in the medium changes from red to orange-yellow or yellow which indicates a drop In pH, decant and discard the medium and replace it with fresh medium.

Results

Results can vary from group to group. Students can quantify viable cells when utilizing the inverted microscope.

14

TRANSPLANTATION TECHNIQUE

On the assumption that the best environment for a cell is the one it encounters in its normal situation in the body, it has been a guiding principle of tissue culture to try reproduce this as closely as possible. Transplantation methods assume that the next best alternative to a tissue's normal environment is a similar environment in another host.

In plants, grafting is a well-established procedure and it is well-known that sometimes the grafts grow even better in another host, a principle which has been used in improving yields from fruit trees. Transplantation in animals has turned out to be much more complicated. The earliest successful attempts were performed in invertebrates and amphibian embryos, and Joest managed to obtain a permanent union between different worms while a litter later Born, Harrison and Morgan were able to transplant tissue in tadpoles. These observations were the precursors of the classical studies on amphibian embryos which form a large part of the literature of experimental embryology. When experiments of this kind were extended to adult animals, particularly of the higher vertebrates, it was found that almost invariably the host rejected the grafted tissue after a few days in a manner which suggested an immune type of reaction. The development of immunity to foreign tissue has proved the main barrier to transplantation in animals.

The relationships involved in the rejection of grafts in higher animals are now clearly recognized. It is well known that tissues can be grafted from one part to another of an individual and, provided the requisite level of technical skill is employed, these *autografts* will survive. Successful grafting can be extended to genetically identical individuals—*uniovular twins* and *inbred animal* strains. Grafts of this kind are called *isografts*. However, when grafts are exchanged between two genetically dissimilar individuals the grafts are rejected after a period of some days (even when the grafts are between two closely related individuals, e.g. binovular twins). Grafts between two individuals of the same species who are not genetically identical are called *homografts* and the rejection of homografts is called the *homograft reaction* or *transplantation immunity*. Grafts between different species are called *heterografts*. Transplantation immunity reactions apply to them also and in much the same way as homograft reactions, provided the tissue can be made to 'take' at all in the first instance.

A characteristic course of events occurs in the rejection of foreign tissue. At first the graft may take and give every appearance of becoming permanently established in the host. After about seven or eight days, however, it becomes invaded by mononuclear cells. The blood and lymphatic supplies stop and the graft begins to deteriorate. Three or four days later the graft is completely destroyed. This reaction has been found to occur in mammals, birds, amphibia and fish. It does not occur in the embryo of these species, however, and the intolerance of foreign tissue seems to develop at about the time of birth or hatching—earlier or later depending on species.

It has been shown that the immunological mechanism of the homograft reaction is quite different from the ordinary antibody type of reaction. In some way the reticulo-endothelial system is involved, especially the *lymphocytes*. These cells seem to possess the ability to recognize foreign tissue when they come in contact with it and to carry this information back to the lymphnodes where antibodies may be produced. Thus, if lymphocytes are unable to reach foreign tissues the homograft reaction is not initiated.

A remarkable feature of this process is that animals may be made tolerant to foreign tissue provided they are inoculated with it before birth (or before the immune mechanism has developed). A naturally occurring example of this is the bovine '*freemartin*' in which binovular twins can be shown to be chimeras (individuals having two genetically distinct species of cells) because of the exchange of some cells *via* a common placental circulation in utero. Experimentally, mice can be rendered tolerant to foreign tissue by inoculation with it at birth. Subsequently they will not reject tissue from the same donor.

The immunological mechanism is sensitive to certain forms of treatment. In particular irradiation or treatment with radiomimetic substances, such as nitrogen mustards, can completely suppress it.

Treatment with cortisone, which is known to have a depressant action on the *reticulo-endothelial system*, can also suppress the homograft reaction. Other substances can produce similar results but no others approach cortisone or x-irradiation in effectiveness.

The above remarks have been concerned with the rejection of normal tissue by hosts of the same species. For some reason which is not understood, certain *neoplastic cells* and tissues do not seem to stimulate the response to the same degree and thus there are some transplantable animal tumours which can be transferred to a variety of hosts. However, transplantable tumours are often carried in the host for a period rather shorter than that required for development of immunity (e.g., *ascites tumours*) and in other cases transmission of the tumour may be due to a virus (e.g., the *Rous sarcoma*).

From the foregoing it can be understood that in order to grow animal tissues in a foreign host the immunological mechanism involved in the homograft reaction must either be circumvented or eliminated. This has been done in the following ways.

1. Transplantation into embryo or very young animals.
2. Transplantation into animals rendered tolerant by treatment with donor cells before the stage of development of transplantation immunity.
3. Transplantation into genetically identical hosts (inbred strains).
4. Transplantation into areas without direct blood supply or lymphatic drainage (anterior chamber of the eye, cornea, brain).
5. Transplantation into artificial chambers from which blood and lymphatic cells are mechanically excluded (by millipore membranes).
6. Suppression of the host's ability to react (by treatment with cortisone and X-rays).
7. Transfer of implanted tissues to a new host before immunity develops (many ascites tumours).

Some cells cultured in vitro can be grown in the animal or passaged through the animal by means of these techniques.

Transplantation into Embryos

The experimental embryologist has performed transplantation experiments between embryos for many years and we know that the success of these depended on the fact that in the embryo the homograft reaction does not develop until a late stage. The techniques do not have a place in this account since they are of a highly specialized

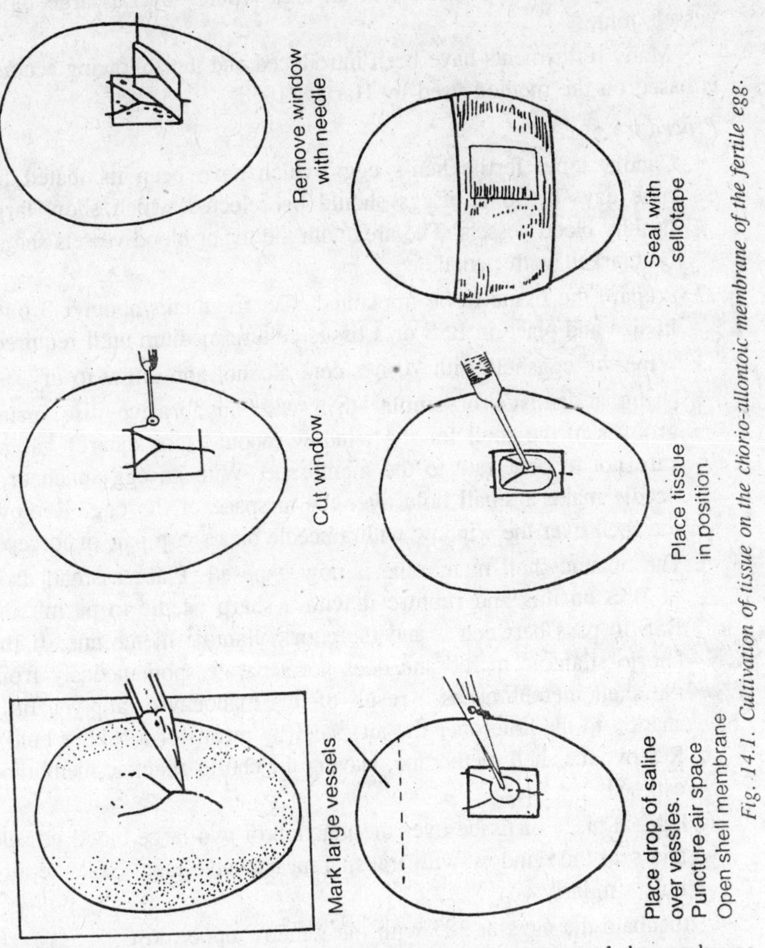

Fig. 14.1. Cultivation of tissue on the chorio-allantoic membrane of the fertile egg.

nature. However, these successful experiments suggested that embryos might provide a general kind of substrate to which animal tissues might be grafted and eventually Murphy, in 1912, was able to demonstrate convincingly the survival of heterologous tumour tissue in the chorio-allantoic membrane of the chick embryo. Since then the fertile hen's egg has been used for the cultivation of both tumour and normal tissue from a variety or adult animals, including man.

Many techniques have been employed. In general, a window is made in the egg shell and the tissue to be planted is placed in the selected site. The yolk sac and the chorio-allantoic membrane have been most frequently used. The preferred technique is implantation on the chorio-allantoic membrane at an area where several large blood vessels join.

Many refinements have been introduced and the following account is based on the method used by Harris (1958).

Procedure

1. Candle some fertile hen's eggs which have been incubated for nine days at 38°C. Eggs should be selected which show large healthy blood vessels. The area with the major blood vessels should be marked with pencil.
2. Prepare the tissue to be implanted. Cut fragments about 1-3 mm. in size and place in BSS or a tissue culture medium until required.
3. Wipe the eggshell with 70 per cent alcohol and allow to dry.
4. Using a dentist's or similar drill with an abrasive disc, grind grooves in the shell for the window (about 1 cm square), taking care not to penetrate to the membrane. With an egg-punch or a needle make a small hole over the airspace of the egg. Remove the shell over the window with a needle or a sharp pair of forceps.
5. The opaque shell membrane is now exposed. Place a small drop of BSS on this and rupture it with a sharp needle to permit the BSS to pass between it and the chorioallantoic membrane. If the chorio-allantoic membrane does not separate spontaneously from the shell membrane as a result of this manoeuvre, apply a little suction to the hole over the airspace (by means of a rubber bulb). Remove the shell membrane, leaving the chorio-allantoic membrane exposed.
6. Place a piece of tissue over the junction of two large blood vessels and seal the window with transparent adhesive tape. Seal the air-hole similarly.
7. Incubate the eggs at 38° with the window uppermost.

After a period of 3-4 days the implanted tissue can be examined by removing the tape over the window. If it is desired to observe the implant continuously good visualization can be obtained by sealing a small coverslip to the eggshell with molten beeswax. Water vapour usually condenses on this and it is necessary to clear it by applying a warm object to the glass prior to examination. An auroscope is useful for this purpose.

Murphy was able to demonstrate growth in the chick chorioallantoic membrane of the Jensen rat sarcoma, the Ehrlich carcinoma, a chondroma of the mouse, a mammary carcinoma of the mouse, the

Flexner-Jobling rat adenocarcinoma and a human sarcoma as well as embryonic tissue of chicken, mouse and rat. Many other workers since then have demonstrated the growth of a variety of human tumours, normal tissue, such as skin and lines of tumour cells, some of which have also been cultured *in vitro*.

Transplantation into Tolerant Chimeras

At the present time this technique has been used only in studies to investigate the nature of the homograft reaction.

Transplantation into Genetically Similar Hosts

Many experimental tumours are maintained by serial transfer (of solid implants or by injection of tissue minces) in inbred animals of the strain from which the tumour originally arose. The techniques are simple. Implants are introduced either by a trocar and cannula or through a small incision into a suitable site. Minces are prepared by a Latapie mincer or similar instrument and injected through a wide-bore needle directly, usually subcutaneously or intramuscularly. No special methods are required in most cases.

Transplantation into Non-vascular Areas

The sites most commonly employed for this purpose are the anterior chamber of the eye and the brain. Attempts to transfer heterologous tissue to the anterior chamber of the eye were made as early as 1884 and at regular intervals thereafter, but at first they were almost uniformly unsuccessful. Henger and Keysser independently reported successful heterotransplantation experiments in 1913 and very many workers have been successful since. Current techniques have been developed mainly by Greene.

Procedure for Anterior Eye Chamber Implantation

1. Prepare the tissue to be transplanted. Suitable implants are about 1 mm. in size.
2. Anaesthetize the animal. For large animals, such as rabbits, Greene uses a local anaesthetic, such as cocaine, applied to the cornea. A general anaesthetic may be used instead. It is essential for small animals, which can be placed in a jar containing some swabs soaked in ether. Nembutal injected intramuscularly, is generally used for larger animals.
3. With the left finger and thumb hold the eyelids open and, by gentle pressure, cause the eye to protrude. Using a double-edged corneal knife open the anterior chamber near the superior border of the corneoscleral junction with a quick jab directing the knife

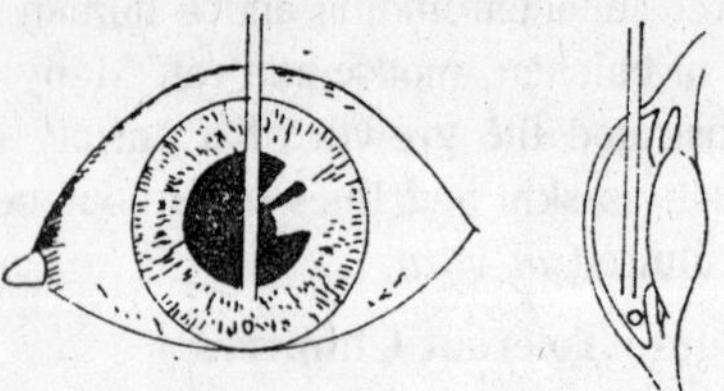

Fig. 14.2. Introduction of tissue into the anterior chamber of the eye.

blade slightly forward to avoid cutting the iris. (Single-edged knives are unsatisfactory).

4. Introduce the tissue through a small cannula (fitted with a stylet or plunger to permit extrusion of the fragment).
5. After withdrawing it, use the side of the cannula to gently massage the tissue down until it is wedged in the inferior angle of the iris.

Certain soft tissues, such as brain, are liable to escape through the corneal incision, particularly in small animals. In cases of this sort Greene recommends that the initial incision should intentionally be carried back through the iris. The cannula should then be inserted into the posterior part of the anterior chamber (behind the iris) and brought forward through the pupil into the lower part of the chamber. On withdrawing the cannula the fragment is then retained at the border of the pupil.

Procedure for Brain Implantation

1. Prepare the implants as before.
2. Anaesthetize the animal. Shave and wash the skin over the upper anterior portion of the parietal bone. Make a small incision and retract the skin to expose the skull.
3. Drill a small hole through the skull, just large enough to take the inoculating cannula (for large animals 1-5 mm. bore; for small animals an adapted hypodermic needle, 20 bore). In mice the hole can be drilled by rotating a scalpel blade. For large animals a drill is necessary and an ordinary metal drilling bit can be used. As soon as the skull has been penetrated the drill must be withdrawn to prevent damage to the brain. This point can be recognized by the sensation of yielding transmitted to the hand.
4. Insert the implant through the cannula, making sure it is expelled into the brain and retained there.
5. Approximate the edges of the skin incision and seal with a drop of collodion or a clip. This technique has been used successfully for rabbits, guineapigs, rats and mice.

Diffusion Chambers

Rezzesi and also Bisceglie in the early 1930s developed methods for short-term cultivation of tissues in collodion bags within the peritoneal cavity of small laboratory animals. Algire and his associates developed the use of millipore membrane chambers for the same purpose. Type HA or Type AA millipore filters can be used. The AA filter excludes particles larger than 0.8 μ and is quite satisfactory.

A diffusion chamber for intraperitoneal implantation is prepared from two rings of perspex and two discs of millipore membrane as follows. The two perspex rings are cut from 5 mm sheet. One is made with an outside diameter of 17.5 mm and an inside diameter of 14 mm., the other with an outside diameter of 13.9 mm, and an inside diameter of 10 mm. A disc 17.5 mm in diameter and another, 14 mm in diameter, are cut from a small piece of millipore membrane. The larger disc is stuck to the larger ring with 1 per cent solution of perspex in acetone. The smaller disc is similarly stuck to the smaller ring.

The tissue is cut to form a small explant about 0.5 mm in size. The surface of the membrane in the larger part of the chamber is moistened with a drop of tissue culture medium and the tissue is placed in it. The inner ring is then placed in position and the junction between outer and inner rings is sealed with the minimum amount of perspex solution.

On removal of the chamber from the animal it can be placed in fixative and stained intact. After dehydration and clearing with xylol the chamber can be taken apart.

Algire also developed an observation chamber on the same principle. In this chamber one side consisted of a glass coverslip and the other of a millipore membrane. The chamber was inserted into a fold of skin on the back of the animal so that continuous observations, by means of a microscope if desired, could be carried out.

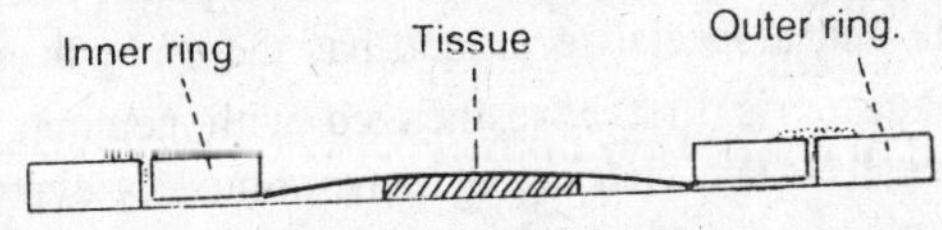

Fig. 14.3. Cross-section of an Algire diffusion chamber.

Transplantation to Irradiated and Cortisone-Treated Animals

It has already been mentioned that the host's immune response can be suppressed in a number of ways and that the best have proved to be irradiation with X-rays and treatment with cortisone, either separately or combined. The development of these techniques has been due mainly to the recent work of Toolan although Murphy demonstrates as long ago as 1914 that irradiation with X-rays permitted hetero-transplantation.

Several kinds of animals have been used as hosts for transplanted tumour tissue. The mouse and rat usually require combined treatment with irradiation and cortisone, but the hamster gives satisfactory results when treated with cortisone alone.

Conditioned rats

Young (week-old) rats are recommended. 1-4 days before inoculation they are subjected to about 150 r of x-irradiation. The inoculum, consisting of 0.5 1 ml. of mince or suspension containing 0.25-0.5 g. tissue, is injected into the flank. At the same time 3 mg. cortisone acetate is administered subcutaneously. This is repeated on alternate days for a total of four injections. The tumours are usually ready for harvesting between the 9th and 12th day.

Conditioned mice

The technique is essentially similar. The animals are given 50-150 r (usually 100 r) total body irradiation 1-4 days before inoculation. An inoculum of about 0.5 ml. of tissue suspension is made either subcutaneously into the flank or, preferably, intramuscularly into the thigh. 1.5 mg. cortisone acetate is injected at the same time and 1 mg. on alternate days for a total of four injections. The tumour may be harvested in 7-12 days.

Conditioned hamsters

The Syrian hamster can be adequately conditioned with cortisone alone. Animals weighing 50-60 g. receive 3 mg. cortisone acetate at the time of inoculation with the tumour and the same dose twice in the following week. They can be inoculated in the same way as rats and mice. However, in the hamster the cheek-pouch offers a particularly suitable site and a special technique has therefore been developed.

The hamster is first anaesthetized with nembutal, administered intraperitoneally. The cheek-pouch membrane is everted and pinned down to a piece of perspex. It is then washed with a mild antiseptic (acriflavine, zephiran). A small puncture is made through the cheek-pouch membrane and tissue fragments are inoculated by means of a small cannula. Alternatively, a cell or tissue mince or suspension may be injected. The cheek-pouch is returned into the mouth.

Fig. 14.4. Introduction of tissue into the hamster chick-pouch.

Besides human tumours, embryonic tissues of a number of species and some strains of cultured cells have been grown by this technique.

Ascites Tumours

The term ascites tumour is applied to certain transplantable tumours of animals which have been made to grow as suspensions of free cells in an exudate within peritoneal cavity. They are of special interest since they behave in very many ways exactly like explanted cell cultures and some of them can be cultured *in vitro*. This permits cells to be alternated between *in vitro* and *in vivo* conditions with great ease.

Hesse, in 1927, demonstrated first that the Flexner-Jobling rat carcinoma could be transmitted by the injection of ascitic fluid which had developed during the growth of the tumour in the peritoneal cavity. Koch repeated and confirmed Hesse's observation the following year. However, the use of ascites tumours in experimental work did not become established until Loewenthal and Jahn, collaborators of Koch's, found in 1932 that after repeated trials they were able to produce an ascites tumour using the Ehrlich carcinoma. Thereafter this tumour became widely used and attempts were made to produce ascites fumours from other material. Since the original work a very large number of these have been developed, mainly in mice and rats. Some are able to grow in many strains while others will grow only in the strain of origin.

Although attempts were made from the beginning to grow these tumours in tissue culture, it proved surprisingly difficult to keep them alive and only recently has any real success been achieved.

Maintenance of ascites tumours

Most ascites tumours are extremely easy to maintain provided a suitable strain of mice of rats is bred for the purpose. Normally, mice with a mature tumour are killed a week after inoculation. The tumour (in suspension in ascites fluid) is removed from the abdominal cavity of infected animals by means of a syringe and needle and reinoculated into new hosts. Details of the procedure as used for the Ehrlich ascites tumour are as follows. Mice containing a mature tumour are put into an ether jar to be anaesthetized. They are then removed one by one and killed by decapitation or dislocation of the neck. The skin of the abdomen is liberally swabbed with alcohol and a syringe with a No. 18 gauge needle is used to extract the cell suspension. Sometimes, difficulty is found in withdrawing the fluid due to the gut being drawn against the end of the needle. Occasionally, it is easier, therefore, to open up the abdomen and remove the ascites fluid by means of a Pasteur pipette. Alternatively, the skin can be reflected to reveal the thin muscular abdominal wall. In mice this is quite transparent and the aspirating needle can be directly observed. The tip of the needle can be used to hold up a small cone of abdominal wall and from this viscera-free area the tumour suspension can be aspirated. On withdrawal of the ascites fluid it is transferred to a test-tube. When all the donor mice have been handled in this way, the mice to be inoculated are then placed in the anaesthetizing jar and lightly anaesthetized. They are removed one by one, the abdominal skin is swabbed with alcohol and the tumour suspension is inoculated direct into the abdomen by means of a syringe and needle. As a general rule, the amount of inoculum varies from 0.1 to 0.25 ml. depending on the rate of growth of the tumour.

Removal of cells for culture in vitro

In the above procedure it is desirable to maintain relatively aseptic conditions but strict asepsis is not essential because the mouse peritoneal cavity has a high resistance to infection. However, if the cells are to be cultured *in vitro*, it is necessary to take somewhat greater care. It is recommended that the skin of the mouse should be everted in the manner described. This exposes a sterile area of abdominal wall through which a needle can be inserted to withdraw some of the cell suspension.

Ascites tumour cells occur in ascites fluid in concentrations varying from about 25 million to more than 100 million per ml. of fluid. It is therefore usually necessary to dilute them very considerably before trying to culture them in the usual media. Otherwise it is exhausted very quickly and the cells die before they have an opportunity to grow. It should be noted that at least 10 per cent, of the cells in the exudate are likely to be cells derived from the peritoneal lining or phagocytes which have entered the peritoneal cavity from the blood.

This factor must therefore be allowed for in experiments. Alternatively, special measures may be taken to remove them. Most ascites tumours tend to grow in suspension *in vitro* and do not settle down on the glass of culture vessels. It is possible therefore, to remove unwanted cells simply by permitting them to stick to the glass of a culture vessel. The supernatant containing the suspended ascites cells can then be pipetted off. This can be done within a matter of hours of removal from the animal.

Preservation, Storage and Transportation of Living Tissues and Cells

Now that very many cell-strains have been isolated, one of the major problems confronting some laboratories is the maintenance of the large number of cultures necessary to carry several strains simultaneously. A further difficulty is the tendency for cell-types to change after they have been kept in a state of rapid proliferation for some years. Similar problems have been encountered in the handling of micro-organisms and the answers which have been suggested for the maintenance and storage of cells are very similar to those which have been adopted by bacteriologists. All the methods depend on the maintenance of the cells at reduced temperatures but there are some differences in technique depending on the temperature used.

Cells can be maintained at slightly reduced temperatures, refrigerator temperatures or very low temperatures (–70°C). Room temperature and refrigerator storage are sometimes used for short-term maintenance but for long-term storage very low temperature are necessary. In the deep-frozen state metabolism is, or course, suspended completely.

Maintenance at Slightly Reduced Temperatures

Most mammalian cells and tissue fragments will survive indefinitely at 30°C provided the medium is renewed when required. By this means the frequency of feeding can be greatly reduce. Thus, compared with 37°C the intervals between feeding can be about three times as long.

The temperature can be still further reduced and many cells will survive unharmed at about 20°. At this temperature very infrequent renewal of medium is required, although the cells rarely continue to behave quite normally if their normal incubation temperature is of the order of 37°C. They tend to round up and many leave the glass and sometimes they become packed with fat droplets. However, if the medium is renewed and the temperature raised to normal once more the cells very rapidly return to normal after a short lag period.

Maintenance at Refrigerator Temperature

Some reports have appeared concerning the storage of tissue in the refrigerator (2-6°). It would appear that sheets of cells and single cells do not survive well for more than a few days in these conditions but that tissue fragments will survive for several weeks in a nutrient medium. This may be due to the high oxygen tension which can develop in the medium. The main practical use of this method is in the preservation of surviving tissue before explanation, usually only for a few days. The tissue should be cut into pieces of a few cubic millimetres and stored in a nutrient medium.

Preservation by Freezing

In several fields advantage has been taken of the observation that deep-frozen tissue may remain viable. The principle was applied to the storage of viruses and bacteria but when it was applied to animal cells they did not survive. At first this was thought to be due to laceration of the cell membranes by ice crystals but more recent evidence suggests that the cause may be osmotic changes which give rise to irreversible changes in lipoprotein complexes resulting in splitting of membranes within the cell. In any event the answer to the preservation of living animal cells proved to be the addition of a substance such as glycerol or ethylene glycol to the medium and slow freezing. The technique was worked out in some detail by Smith who demonstrated the survival of ovarian granulosa cells after deep freezing. Scherer and Hoogasian demonstrated its effectiveness with some stock lines of cultures cells and since then it has been widely applied.

Some differences of opinion exist as to the best conditions for survival of frozen cells. Some authors maintain that the rate of freezing is unimportant while others seem to have shown convincingly that slow freezing gives better results. In the circumstances it would seem to be safer to freeze slowly, over a period of 40-60 minutes. All authorities are agreed that it is better to store cells at the lowest temperature possible and -70°C is usually employed. They usually die if stored for any length of time at a temperature higher than -20°C. The reason is not clear but it is probably due to the fact that certain proteins may be denatured even at this temperature. Glycerol is added to the medium frequently at a level of 15 per cent, though different concentrations are quoted.

The following principles summarize generally accepted current practice.

1. The cells should be in a healthy state before freezing.
2. They should be suspended in growth medium containing glycerol (15 per cent) shortly before freezing, and sealed in a gas-tight ampoule.
3. Freezing to –70° should be controlled over a period of about an hour or preferably longer. A cooling rate of one degree a minute is frequently recommended.
4. Storage temperature should be maintained at –70° or lower.
5. Thawing should be rapid (2-3 minutes).

Cells stored in this way tend to deteriorate slightly but survival times from two to five years have been obtained. It is probably desirable to thaw, culture and refreeze at least once per annum.

15

FORMATION OF PROTOPLASTS

Protoplasts are plant cells whose cell walls have been removed. Recall that plant cells *in vivo* are surrounded by rigid cell walls composed largely of celluloses and cemented together by pectins. The cell wall normally confines the cell, preventing it from bursting as a result of turgor pressure. For some biotechnology and genetic engineering protocols, however, the cell wall is a substantial barrier. Releasing plant cells from the confines of the cell wall allows them to be manipulated by microinjection, electroporation and fusion.

Many techniques for the production of protoplasts have been developed. Originally, tissues were bathed in a solution of high osmotic potential to shrink the cells. Then, cell walls were broken mechanically by cutting or abrasion. Finally, by reducing the osmotic potential of the medium, the cells were made to swell and pop out of the broken cell walls. The yield of viable protoplasts from this type of protocol is generally low, thus, except for some difficult tissues, this strategy is rarely followed. Most methods today depend on the enzymatic breakdown of cell walls first demonstrated by Cocking (1960). Protoplast digestion mixtures generally include several enzymes, the concerted action of which efficiently releases cells from the cell wall. A mixture of cellulase, pectolyase, and macerase will be used in this experiment. Because naked plant cells are fragile, and can burst in solutions of low osmotic potential, mannitol is included in the digestion mixture as an osmoticum. Other ingredients included improve the stability of the protoplasts.

The plant material used in the experiment is red onion. From this plant, you will prepare protoplasts from the bulb (a modified leaf structure). Because not all cells of red onion are pigmented, protoplasts are a mixed population of colored and colorless. After preparing the protoplasts, you will use them to perform a cell fusion experiment. By treating the protoplasts with solutions of polyethylene glycol and calcium, they can be made to fuse to form hybrid cells. These cells will be recognizable under the microscope. Can you think of any possible applications of this procedure in biotechnology?

If your purpose in this experiment were to grow up the hybrid cells or even to regenerate whole plants from them, it would be necessary for you to isolate the protoplasts from sterile (i.e. axenic) tissue, to maintain sterility throughout the protocol, and to return the hybrid cells to culture where they could continue their development. These manipulations are not trivial. In practice, a protoplast fusion experiment of the type you are doing here would be done on a much larger scale and the protoplasts would be treated somewhat differently. First, the protoplasts to be fused would likely originate from different tissues, generally from different genotypes or even different species. The protoplasts would be washed after isolation to remove traces of the *digestive enzymes* used to remove their cell walls. After fusion, they would be plated in media by techniques that encourage them to regenerate cell walls. In this laboratory experiment, the precautions necessary to maintain your protoplasts in long term culture will not be followed, but you should understand how this experiment would be different if it had different goals.

Safety Guidelines

Follow standard laboratory safety practices.

Experimental Outline

Cut up onion tissue

↓

Incubate in digestion mixture (1 hr)

↓

Collect protoplasts, observe, allow to settle on slide (20 min)

↓

Add fusion buffer and incubate (20 min)

↓

Add W10 buffer and observe fusion (20 min)

Materials

Plant material: Red onion. Choose a firm unblemished bulb.

Petri dishes or depression slides: Small diameter (60 mm) disposable dishes or microscope slides with depressions or wells are the container of choice for digesting the plant tissue. Small test tubes will work, but petri dishes offer the advantage that they can be viewed on a microscope so that students can monitor protoplast release.

Microscope: An inverted scope, of the type normally found in tissue culture labs, is the instrument of choice for this use. If such a scope is not available, the normal student scope will work if the experiment is done on a microscope slide. Depending on the magnification available, even a dissecting scope can be used. Since the success of this experiment depends on the students' viewing the results, test available equipment before attempting the lab exercise.

Pasteur pipets

Enzyme solution:

- 1.5% Cellulysin
- 1.5% Macerase
- 0.2% Pectinase
- 0.4 M Mannitol
- 2.0% Glycine

PEG Fusion Buffer:

- 0.3 M glucose
- 66 mM $CaCl_2$
- 40% polyethylene glycol (molecular weight 1400)-pH 6.0

Solution W10:

9 parts solution A:	0.4 M glucose
	66 mM $CaCl_2$
	10% dimethylsulfoxide
1 part solution B:	0.3 M glycine-OH buffer pH 10.5

Pre-lab Preparation

Timetable of events

Prepare protoplasts (1 hr)

Do fusion experiment (45 min)

Filter the enzyme solution through a 0.45μm filter (Gelman) before use. The solution can be stored frozen for several months without significant loss of activity. Also prepare some of the same solution without enzymes for use as "medium" in step 6 of the student's experimental procedures.

Method

1. Using a razor blade or scalpel, cut a small square of tissue (about 1 cm^2) from one of the layers of a red onion. You will notice on the cut edge that the tissue is only red on one surface. Slice the tissue again, layer cake fashion, to expose cells that were buried in the center of the square.
2. Place the half of your tissue containing both red and white cells face down in a small pool (about 0.2-0.5 ml) of enzyme solution. Cover the container in which the tissue is being digested to prevent the enzyme solution from evaporating. Allow the enzyme to work for 1/2 to 1 hour. You will want to examine the digestion periodically during this time to monitor the release of protoplasts. It will help to occasionally agitate the mixture gently.
3. At the end of 1 hour, remove with forceps any remaining bits of undigested tissue. Remove 2 drops of protoplasts to a clean microscope slide. Protoplasts are spherical structures. Do not cover them with a coverslip. Check the slide on the microscope to see that you have both red and colorless protoplasts. The actual number of protoplasts on the slide is not critical, but there should be enough in your sample that they are not difficult to find. Typically, several dozen should be readily visible. If this is not the case, make a new slide. (The protoplasts are actually large enough so that you can see to collect them with a pipet.)
4. Place the slide on the lab bench where it will not be disturbed and allow the protoplasts to settle onto the glass about 15-20 minutes.
5. Add 1 drop of PEG Fusion Buffer to the protoplasts. Check the slide on the microscope at this point. You should see the protoplasts clumping together. Wait 15 minutes, then carefully remove the liquid from the slide with a pipet. The protoplasts should remain behind.
6. Add 2 drops of solution W10 to the slide. Check the slide on the microscope periodically over the next 20 minutes during which fusion may begin to occur. At the end of this time, if no fusion has occurred, add several drops of medium to the slide and view again.

Results

1. Draw a picture showing: colorless and red protoplasts, fusion products.
2. Label the following: plasma membrane, vacuole, nucleus.

3. How can you distinguish between fused and non-fused protoplasts?
4. If a problem arises, it is likely to be that protoplast yield or yield of fused protoplasts is low. Even under these circumstances, positive results from at least a portion of the class should be forthcoming from the experiment.
5. Since handling the protoplasts under sterile conditions, as would be done in a research lab, presents difficulties for the teaching lab, the experimental protocol ends with the production of fusion products. Students should understand how the experiment would be performed differently if the goals of the experiment were to maintain the hybrids in culture. This is discussed in the student guide.
6. The plant material used in the experiment, red onion, was chosen because it is readily available and produces a good yield of protoplasts within the time limitations imposed by a laboratory class. The other benefit of using red onion for this experiment is that digestion of the red onion tissue gives a mixed population of cells, red and colorless. This means that for the fusion experiment included in the module, it is not necessary to prepare two different plant tissues. Other plant materials may perform equally well, but substitutions should be tested before class since all plant tissues do not respond equally to this mixture of enzymes and osmoticum. Some tissues will take from a few hours to overnight to release a sufficient number of protoplasts for the fusion experiment. If desired, protoplasts may be prepared before class and used for the fusion experiment alone during the laboratory period.

16

Culture of Specific Cell Types

It will be apparent from the discussion in previous chapters that the expression of specialized functions in culture is controlled by the nutritional constitution of the medium, the presence of hormones and other inducer or repressor substances, and the interaction of the cells with the substrate and other cells. Reviews of specialized culture techniques for specific cell types can be found Barnes et. al., (1984a-d) in Jakoby and Patsan (1979), Kruse and Patterson (1973), Willmer (1965), Defendi (1964), Ursprung (1968), and Sato (1978, 1981). My main purpose here is to present an outline of some of the techniques that are available for culturing different tissue or cell types and to exemplify the diversity of cell types that can be cultured. It is useful to classify these anatomically, although there is considerable overlap in the techniques used.

A number of specialized procedures have now been devised and some representative examples have been contributed by experts in each area. It is assumed in the following protocols that the basic prerequisites of the cell biology laboratory, will be available. Consequently items such as inverted microscopes, bench centrifuges, and waterbaths will not be reiterated in each Materials section.

Epithelial Cells

Epithelial cells are often responsible for the recognized functions of an organ, e.g., controlled absorption in kidney and gut, secretion in liver and pancreas, and gas exchange in lung. They are also of interest as models of differentiation and stem cell kinetics (e.g., *epidermal keratinocytes*) and are amongst the principal tissues where the common cancers arise. Consequently, culture of various epithelial cells has been a focus of attention for many years. The major problem in the culture of pure epithelium has been the overgrowth of the culture by stromal cells such as the connective tissue fibroblasts and vascular endothelium. Most of the variations in techniques are aimed at preventing this, by nutritional manipulation of the medium or alterations in the culture substrate.

Factors contained in serum, many of them derived from platelets, have a strong mutogenic effect on fibroblasts, and tend to inhibit epithelial proliferation by inducing terminal differentiation. Consequently, one of the most significant development in the isolation and propagation of specialized cell cultures has been the development of selective, serum-free media, supplemented with specific growth factors as appropriate.

Isolation of epithelial cells from donor tissue is best performed with collagenase as this disperses the stroma but leaves the epithelial cells in small clusters, which favours their subsequent survival.

Epidermis

Rheinwald and Green (1975) showed that murine and human epidermal kerationcytes could be cultured selectively on feeder layers of irradiated 3T3 cells and could mature to form differentiated squames. *Basal cell carcinoma* can also be cultured by this method. Alteration of the constituents of the medium enabled Peehl and Ham (1980) and Tsao et al. (1982) to culture keratinocytes from human foreskin selectively without feeder layers or serum and others have shown that reductions in pH, Ca^{2+} and temperature may all contribute to improved selective growth of epidermal keratinocytes. Addition to hydrocortisone (10μg/ml), 10^{10} M cholera toxin, or 10^{-6} M isoprenaline (*isoproterenol*) and epidermal growth factor (10 ng/ml) to the medium has made continued serial subculture possible over many cell generations. When mouse-sublingual epidermal cultures were grown on collagen rafts at the gas-liquid interface, complete histological maturation was possible.

The following protocol for the culture of epidermal cells has been contributed by Norbert E. Fusenig, Institute of Biochemistry, German Cancer Research Center.

Principle

Protease digestion of skin samples leads to a separation between epidermis and dermis with the split level being partly beneath and partly above the basal cell layer depending on tissue sample and incubation temperature. Single cells are obtained by mechanical dispersion.

Keratinocytes of animal and human skin have been grown in primary culture and for limited numbers of subcultures using a variety of substrate, culture media, and additives including feeder layers. Cultures tend to stratify forming differentiating multilayers when grown at normal (1.4 mM) Ca^{++} concentrations, but stay essentially as "undifferentiated" monolayers in media with low (below 0.1mM) Ca^{++} concentrations. Maintenance at a high cell density or at low density in the presence of confluent feeder cells helps to reduce fibroblast contamination.

Outline

Separate dermis from epidermis by cold trypsin digestion, collect Keratinocytes by pipetting epidermal and scraping dermal layers, and propagate in low Ca^{++} medium.

Materials

Tissue culture media:

1. Eagle's MEM with a 4 × concentration of vitamins and amino acids (essential and nonessential) (4 × MEM) and 15% heat-inactivated fetal calf serum (FCS)
2. The same 4 × MEM medium with Ca^{++} -free HBSS, Chelex treated FCS and the Ca^{++} concentration adjusted by added $CaCl_2$ or normal FCS to a final concentration of 0.08 and 0.22 mM for mouse and human cells, respectively.
3. FAD-medium consisting of a mixture of 1 part medium F-12 and 3 parts of DMEM enriched with adenine (1.8×10^{-4}M) cholera toxin (10^{-10}M), EGF (10 ng/ml), hydrocortisone (0.4 mg/ml), and 5% FCS. All media contain antibiotics (*penicillin* 100 U/ml and streptomycin 50 mg/ml).

trypsin (1: 250)

HBSS

DBSS (HBSS + antibiotics)

PBSA

100 mm bacteriological grade petri dishes

scalpels, curved forceps

Protocol

1. Wash 5-10 times in DBSS, changing the instruments at least once during rinsing.

2. To provide better and consistently good access of the enzyme to the epithelial-mesenchymal border zone, the skin specimens have to be dissected into pieces of equal size (approximately 1 × 2 cm) with scalpels. These specimens are placed epidermis-side-down into dry bacterial plastic petri dishes and irrigated with a few drops of PBS. The subcutaneous and lower dermal tissue is cut off as much as possible with curved scissors.
3. The tissue samples are rinsed again (five to ten times) in sterile Ca^{++} - and Mg^{++} - free PBS and floated on ice-cold 0.2% crude trypsin (1:250) in Ca^{++} — and Mg^{++} — free PBS (pH 7.2). Usually five to eight pieces are floated on 10-15 ml trypsin in 100-mm (diameter) plastic petri dishes which are placed in a 4°C refrigerator under sterile conditions for 15 to 48 hr. The pH of the trypsin has to be controlled by added indicator dye (e.g., phenol red) to prevent pH shift leading to altered enzyme activity and loss in cell viability.

   Alternatively, tissue strips are floated on 0.2% trypsin for 1 to 3 hr at 37°C in the incubator. At this temperature incubation time has to be carefully controlled to prevent cell damage by the protease.
4. When the first detachment of epidermis is visible in the cut edges of skin samples, the pieces are placed (dermis-side down) in 100-mm plastic petri dishes irrigated with 5 ml complete culture medium including serum. With two fine curved forceps the epidermis is gently peeled off and pooled in a 50-ml centrifuge tube containing 20 ml complete culture medium. Viable keratinocytes are detached from the epidermal parts by vigorous pipetting and sieving through nylon gauze (100-μm mesh).
5. The remaining dermal part is gently scraped with curved forceps on its epidermal (upper surface) to remove loosely attached basal cells. The isolated cells from dermal and epidermal parts are combined, washed twice in culture medium by centrifugation at 100 g for 10 min, and counted for total and viable (trypan blue excluding) cells.
6. Cells are plated at 37°C in medium (1) or (3) at 1.5 × 10^5 cells/cm^2 and left undisturbed for 1 to 3 days, depending on species and isolation procedure, for attachment and spreading.
7. When cells have attached, cultures are extensively rinsed with medium to eliminate nonattached dead and differentiated cells and cultivation is either continued in medium (1) or (3) for long-term

growth of stratified cultures or changed to medium (2) and continued at low Ca^{++} concentration.

8. The shift to low Ca^{++} culture medium (2) is done by rinsing 1-3-day-old primary cultures twice with low Ca^{++} medium and further incubating cultures in this medium. Remaining cell aggregates or differentiated cells sticking to the monolayer detach in this medium within a few days so that subsequent cultures are essentially monolayer.
9. For subcultivation, cultures in medium (1) and (3) are first incubated in 0.05-0.1% EDTA to initiate cell detachment for 10-30 min until cells start to round up, visible by the enlargement of intercellular spaces. Final detachment is achieved by incubation in 0.1% trypsin and 0.05% EDTA and pipetting; the latter procedure alone is sufficient for cultures in medium (2).
10. Human and mouse cells can be grown in fall three media for several months. While mouse cells can only be subcultured once or twice human cells can be passaged for four and seven times in media (2) and (3), respectively.
11. Cultured cells have to be characterized for their epidermal (epithelial) nature to exclude contamination by mesenchymal cells. This is best achieved using cytokeratin-specific antibodies for epithelium and antivimentin for mesenchyme-derived cells. Further identification criteria for other epidermis-derived cells in keratinocyte cultures are available by histochemical and immunocytochemical methods.

Variations

Keratinocytes can also be grown at clonal density co-cultured with x-irradiated 3T3 cells, at reduced Ca^{++} concentrations with fibroblast conditioned medium and in defined serum-free medium.

In order to provide more in vivo-like growth conditions, "*organotypic*" culture systems have been developed by seeding the cells on collagen gels pieces of dermis and lifting these supports to the air-medium interface. Under these improved growth conditions keratinocytes express many aspects of growth and differentiation of the epidermis *in vivo* which are less absent or pronounced in submerged cultures on plastic.

Growth under mesenchymal influence can be provided by transplantation of cell suspensions or intact cultures *in vivo* or by recombining cultures with mesenchyme *in vitro*. This leads to an almost complete expression of growth and differentiation characteristics normal epidermis.

Alternative methods have been described for cell isolation in large quantities and further purification with Ficoll density gradients or with Percoll gradients, methods which are particularly useful for newborn rodent epidermis.

Breast

Milk and reduction mammoplasty are suitable sources of normal ductal epithelium from breast, the first giving purer cultures of epithelial cells. Growth on confluent feeder layers of fetal human intestine repress stromal contamination with both normal and malignant tissue, and optimization of the medium enables serial passage and cloning of the epithelial cells. Cultivation in collagen gel allows three-dimensional structures to form; these correlate well with the histology of the original donor tissue.

As with epidermis, cholera toxin and EGF stimulate the growth of epithelioid cells from breast *in vitro*.

The hormonal picture is more complex. Many epithelial cells survive better with insulin added to the culture (1-10 IU/ml) in addition to *hydrocortisone* (~ 10^{-8} M). The differentiation of acinar breast epithelium in organ culture requires hydrocortisone, insulin, and prolactin and requirements for estrogen, progesterone, and growth hormone have been demonstrated in cell culture.

Principle

Early lactation and postweaning milk, which give the highest cell yield, contain clumps of epithelium which can proliferate in culture. Primary cultures, grown in hormone supplemented human serum containing medium, give cell lines of limited lifespan but clonogenic. These are eventually overtaken by nonepithelial "late milk" cells.

Outline

Cells centrifuged from early lactation milk are grown in the presence of endogenous macrophages in an enriched medium. These may be subcultured with a mixed protease chelating solution.

Materials

Nunc plastic dishes

Sterile universals or 20-50 ml centrifuge tubes

Medium RPMI 1640 (Gibco)

Fetal calf serum (FCS, Flow Labs)

Human serum (HuS outdate pooled serum from blood blanks; Australian antigen negative)

Stock solution

Insulin (Sigma), 1 mg/ml in 6 mM HCl

Hydrocortisone, 0.5 mg/ml in physiological saline

Cholera toxin (Schwartz-Mann), 50 Mg/ml in physiological saline

Serum and stock solutions of insulin, hydrocorstisone, cholera toxin, pancreatin, and trypsin should be kept –20°C.

Trypsinization solution (TEGPED):

10 ml 0.5% EGTA (Sigma, ethylene glycol-bis (β-aminoethylether) N'N' tetraacetic acid) in PBSA

4 ml 0.02% EDTA (Sigma, diaminothene tetra-acetic acid) in PBSA

4 ml 0.2% trypsin (Difco) in Hanks' balanced salt solution (HBSS)

2 ml 1.0% pancreatin (Difco) in HBSS

Growth medium

RPMI 1640 containing 15% FCS, 10% HuS, Cholera toxin 50 ng/ml, hydrocortisone 0.5 Mg/ml, insulin 1 μg/ml.

Protocol

Milk collection

Milk (2-7 days postpartum) can best be collected on hospital wards. The breast is swabbed with sterile H_2O and the milk manually expressed into a sterile container. Five to 20 ml are usually obtained per patient. The milks are pooled and diluted 1:1 with RPMI 1640 medium to facilitate centrifugation.

Primary cultures

1. Spin diluted milk at 600-1000 g for 20 min. Carefully remove supernatant leaving some liquid so as not to disturb the pellet.
2. Wash the pelleted cells two to four times with RPMI containing 5% FCS until supernatant is not turbid.
3. Resuspend the packed cell volume in growth medium and plate 50 μl packed cells in 5-cm dishes (Nunc), in 6 ml growth medium. Incubate at 37°C in 5% CO_2.
4. Change medium after 3-5 days and thereafter twice weekly. Colonies appear around 6-8 days and expand to push off the milk macrophages, which initially act as feeders.

Subculture of milk cells

Incubate in TEGPED (1.5 ml per 55-cm plate) at 37°C for 5-15 min, depending on the age of the culture, to produce a single-cell suspension which can be diluted one third in fresh medium for replating (after washing free of enzyme mixture).

Modifications of the technique

It is convenient to use the macrophages which are already present in the milk as feeders, and these are gradually lost as the epithelial colonies expand. However, macrophages can be removed by absorption to glass and in that case other feeders must be added. Irradiated or mitomycin-treated 3T6 cells show the best growth-promoting activity. Analogues of cyclic AMP can be used to replace the *cholera toxin*, although this is not possible with macrophage feeders, which are killed by the analogues.

Uses and application of milk epithelial cell cultures

Milk cultures provides cells from the fully functioning gland and allow definition of phenotypes by immunological markers. They have been successfully transformed by SV40 virus and provide an important source of normal cells for comparison with breast cancer cell lines and for transfection with oncogenes.

Cervix

A modification of the epidermal culture technique can be used for the propagation of cervical epithelium.

The following protocol for the culture of epithelial cells from cervical biopsy samples has been contributed by Mary Freshney. It is based on a protocol devised by Stanley and Parkinson (1979).

Principle

Epithelial cells cultures maybe established from biopsies of the cervix by culturing cells on a feeder layer of irradiated 3T3 cells in the presence of growth factors.

Outline

Punch biopsy samples of tissue are diced and after initial growth are seeded onto irradiated 3T3 cells at first passage.

Materials

Transport medium

Leibovitz L15 with 10% FCS: 0.5 Mg/ml hydrocortisone, 50 U/ml penicillin, 50 μg/ml streptomycin

Trypsin solution for disaggregation of epithelial cells: 0.1% trypsin + 0.01% EDTA in PBSA

Epidermal growth factor (EGF from BRL) (× 200 stock); 100 μg dissolved in 5 ml sterile water; dilute to 1 μg/ml.

Cholera enterotoxin (Schwartz-Mann) 856011 from Becton Dickinson: mol wt 60,000; make 1,000 × stock solution of 10^{-7} M.

Hydrocortisone

Sigma H4001 (× 1000 stock): weigh 25 mg, dissolve in 3 ml absolute alcohol, add 47 ml water; filter sterilize, aliquot, and keep dark at 4°C.

Growth medium of primary explants

Leibovitz L15 with 10% FCS, 50 U/ml penicillin, 50 Mg/ml streptomycin, 0.05 g/ml hydrocortisone, 10^{-10}M cholera toxin (EGF 5 ng/ml added after explants attach; EGF inhibits plating) tissue culture flasks 25 cm² and 75 cm² growth area.

9-cm petri dishes

Sterile forceps sterile scalpels with Beaver blades number 65-mini

Sterile pipettes

Universal containers or conical centrifuge tubes

Hemocytometer

Irradiation source, e.g., ^{60}Co

Growth medium for 3T3 cells and epithelial cells on irradiated 3T3 cells-MEM + 10% FCS

PE

3T3 Swiss fibroblasts

These cells should be maintained at low density. After trypsinization and F75 should be reseeded at 5 × 10^4-10^5 cells/flask and 175^2-cm seeded at 2 × 10^5 cells/flask. Feed after 3 d and then subculture 3-4 d later when an 75 will contain 3-5 × 10^6 cells and a 175^2-cm flask will contain 15-20 × 10^6 cells.

To Irradiate 3T3 cells

Trypsinize flasks of cells, resuspend in medium-MEM + 10% FCS—and count cells. Irradiate cells in a universal container giving 60 Gy (6,000 rads) from a cobalt source. Inoculate F75 flasks with 1.5 × 106 irradiated cells/flasks and incubate until the cells are attached (From 5 hr to overnight). Replace medium with MEM + 10% FCS + H.C. + CET (don't add EGF until epithelial cells have attached).

Protocol

1. Collect biopsy samples in transport medium and store 1-4 days at 4°C until convenient to set up cultures.
2. Take five 9-cm petri dishes and put 15 ml DBSS into three, 5 ml into one, and leave one empty.
3. Transfer sample with transport medium into empty petri dish.
4. Transfer sample to first dish of BSS and rinse.
5. Transfer sample to second dish. Using two scalpels with Beaver blades number 65-mini, pull away any connective tissue that may be attached to the sample. This is obvious as it is thin and stringy.
6. Transfer sample to third dish of BSS and rinse.
7. Transfer sample to dish with 5 ml DBSS and start chopping. This is easier if done on a dry part of the dish. Cut through the tissue as cleanly as possible without pulling it apart until the pieces are about 2 mm across. Let the BSS wet the pieces.
8. Using a pipette transfer the pieces to a universal container or centrifuge tube. (Wet the inside of the pipette first with DBSS to prevent the pieces from sticking to it). Allow the pieces to settle.
9. Withdraw the supernatant and discard; add 10-15 ml fresh DBSS to container and allow pieces to settle.
10. Repeat once using growth medium.
11. Discard medium and add about 1 ml of fresh medium.
12. Label and date two to four F25 flasks (depending on the number of explants you have) and add 2.5 ml growth medium to each.
13. Add 0.25 ml of medium containing approximately ten pieces to each flask.
14. Incubate at 37°C until some of the explants attach. This may take up to 10 days.
15. Add EGF to 5 ng/ml final concentration. If many cells are growing out from the explants, replace medium with medium containing EGF.
16. Continue to feed the cultures every 3-4 until they become one-half to two-thirds confluent, when they will be ready to be subcultured onto irradiated fibroblasts.
17. To trypsinizine primary cultures, wash with PE. Add 2.5 ml 0.1% trypsin 0.01 EDTA in PBS. Leave for 1 min and remove. Incubate flask at 37°C until cells loosen.
18. Resuspend cells in medium and count using a hemocytometer.
19. Inoculate flasks containing irradiated 3T3 cells with 10^4–10^5 cervical cells/flask.

20. Incubate and feed every 3-4 d until epithelial cells have grown.
21. Subculture epithelial cells on to fresh irradiated 3T3 cells before they become confluent.

Variations

The epithelial cell layer may be separated from the feeder layer by treatment with Dispase II (Protease, Boehringer Mannheim 165 859). Stock solution: 5 g dissolved in 20 ml PBSA (gives a cloudy solution, but sediment will settle out or can be filtered off). Sterilize by filtration, aliquot, and store at –20°C.

When cultures are confluent add Dispase II to 1.2 U/ml (in medium), incubate at 37°C for 1 hr, when the epithelial cells should have rolled up and come off. Remove medium and cells and spin at 100 g for 10 min. Discard medium and wash cells with PE, spin, and discard PE.

Gastrointestinal Tract

Culture of normal gut lining epithelium has not been extensively reported although there are numerous reports in the literature of continuous lines from human *colon carcinoma*. *Colorectal carcinoma* cells plated on confluent feeder layers of FHI from colonies which apparently disappear but reappear 8-10 wk later as nodules in the monolayer. These nodules will increase and can be subcultured with or without feeder layers and, in several cases, have given rise to continuous cell lines, some of the which produce *carcinoembryonic antigen* (CEA) and sialomucin, markers of neoplasia in human colon.

Owens et al. (1974) were able to culture cells from fetal human intestine as a finite cell line. Similar results have been obtained in the author's laboratory, where more vigorous growth was observed in the epithelial cell component of these cultures and fibroblastic cells were eventually diluted out, giving rise to a finite cell line, FHI.

Although no protocol is available for the propagation of normal cell lines form the gastrointestinal tract, at a time of writing it seems probable that the promising approach adopted by Ham and colleagues will lead to a selective medium for gut epithelium in the near future. Such media as MDCB 153 or HLC 9 would seem to be an appropriate starting point.

Liver

One of the major objectives of the 1960s was the culture of cell lines of normal functional liver parenchyma. This has not been fully realized, but there are now many examples of epithelial cells cultured from rat liver. Although these cultures do not express all the properties of liver parenchyma, there is little doubt that the correct lineage of cells may be cultured even if the correct environment cannot be re-created for full functional expression.

The development of the correct conditions for perfusing liver *in situ* with *collagenase* and *hyaluronidase* provided a technique whereby a good viable suspension of liver parenchymal cells could be plated out with high purity and form a viable monolayer, and reports such as those of Malan-Shibley and Type (1981) suggest that some epithelial cultures may be propagated.

Some of the most useful continuous liver cell lines were derived from Reuber H35 and Morris minimal deviation hepatomas of the rate. Induction of *tyrosine aminotransferase* in these cell lines with *dexamethasone* proved to be a valuable model for studying the regulation of enzyme adaptation in mammalian cells.

Pitot and others have demonstrated that, as with epidermis, greater functional expression can be obtained by culturing liver parenchymal cells on free floating collagen sheets. The cells are seeded in medium onto a preformed collagen gel on the base of a petri dish. After the cells have attached, the gel is released from the base of the dish with a bent Pasteur pipette or spatula, allowing it to float freely in the medium. This permits access of nutrients to the cell from above and below. It is possible that diffusion of nutrients and metabolites via the collagen layer is analogous to the situation *in vivo* where epithelial cells usually lie on basement membrane, and may be important in establishing a necessary polarity in the cells.

Principle

Low cytotoxic proteolytic enzymes such as *collagenase*, when perfused into the liver through the vessels and capillaries at an adequate flow rate, will disrupt intercellular junctions, and will digest the connective framework within 15 min if the liver is previously cleared of blood and depleted of Ca^{++} by washing with calcium-free buffer. Hepatocytes are selected from the cell suspension by two or three differential centrifugations.

Outline

Introduce a cannula in the portal vein or a portal branch, wash the liver with a calcium-free buffer (15 min), perfuse with the enzymatic solution (15 min) collect and wash the cells, and count the viable hepatocytes.

Materials

Peristaltic pump (10 to 200 rpm.)

Waterbath

Sterile Tygon tube (ID 3.0 mm; OD 5.0 mm), disposable scalp vein infusion needles, 20 gauge (Dubernard Hospital Laboratory, Bordeaux, France), and sewing thread for cannulation

Chronometer and sterile graduated bottles and petri dishes

Sterile surgical instruments (sharp, straight, and curved scissors and clips) 2 × 1-ml disposable syringes

Heparin (Roche)

Nembutal (Abbot, 5%)

Calcium-free Hepes buffer pH 7.65 : 160.8 mM NaCl; 3.15 mM KCl; 0.7mM $Na_2HPO_4 12H_2O$, 33 mM Hepes-sterilization by filtration on 0.22-μm Millipore filters and storage at 4°C (2 months)

Collagenase (Sigma grade I; Worthington CLS; Boehringer 103578)

Collagenase solution: 0.025% collagenase; 0.075% $CaCl_2 2H_2O$ in calcium-free Hepes buffer pH 7.65-preparation and sterilization by filtration just before use.

L_{15} Leibovitz medium.

Protocol for Isolation of rat hepatocytes

1. Warm the washing Hepes buffer and collagenase solution in waterbath (usually approximately 38-39°C to achieve 37°C in the liver). Oxygenation is not necessary.
2. Set the pump flow rate at 30 ml/min.
3. Anesthetize the rate (180-200g) by intraperitoneal injection of nembutal (100 μl/100 g) and inject heparin into the femoral vein (1,000 IU).
4. Open the abdomen, place a loosely tied ligature around the portal vein approximately 5 mm from the liver, insert the cannula up to the liver, and ligate.
5. Rapidly incise the subhepatic vessels to avoid excess pressure and start the perfusion with 500 ml calcium-free Hepes buffer, at a flow rate of 30 ml/min; verify that the liver whitens within a few sec.
6. Perfuse 300 ml of the collagenase solution at a flow rate of 15 ml/min for 20 min. The liver becomes swollen.
7. Remove the liver and wash it with Hepes buffer; after disrupting the Glisson capsule, disperse the cells in 100 ml L_{15} Leibovitz medium.

8. Filter through two-layer gauze or 60-80 μm nylon mesh, allow the viable cells to sediment for 20 min (usually at room temperature), and discard supertant (60 ml) containing debris and dead cells.
9. Wash three times by centrifugations (50 g for 40 sec) to remove collagenase, damaged cells, and non-parenchymal cells.
10. Collect the hepatocytes in Ham's F12 or Williams'E medium enriched with 0.2% blue exclusion test (0.2% w/v, Sigma) and 10 mg/ml bovine insulin (80-100 ml).

Analysis

Determine cell yield and viability by the well preserved-refringent shape or the trypan blue exclusion test (0.2% w/v) (usually 400 to 600 $\times 10^6$ viable cells with a viability of more than 95%).

Isolation of hepatocytes from species

The basic two-step perfusion procedure can be used for obtaining hepatocytes from various rodents, including mouse, rabbit guinea pig, or woodchuck by adapting the volume and the flow rate of the perfused solutions to the size of the liver. The technique has been adapted for the human liver by perfusing a portion of the whole liver (usually at 1.51 Hepes buffer and 11 collagenase solution 70 and 30 ml/min respectively) or biopsies (15 to 30 ml/min depending on the size). A complete isolation into a single cell suspension can be obtained by an additional collagenase incubation at 37°C under gentle stirring for 10 to 20 min (especially for human liver). Fish hepatocytes can be obtained by cannulating the intestinal vein and incising the heart to avoid excess pressure. Perfusion is performed at room temperature at a flow rate of 12 ml/min.

Application

Isolated parenchymal cell can be maintained in suspensions for 4-6 and used for short-term experiments. They survive for a few days when seeded in the nutrient medium supplemented with 10^6 M dexamethasone on plastic culture dishes (7×10^5 viable cells/ml). A survival of several weeks is obtained by seeding the cells onto a biomatrix. However, they rapidly lose their specific functions. A high stability (2 months) can be obtained by co culturing hepatocytes with rat liver epithelial cells presumed to derive from primitive biliary cells.

Pancreas

There has not been the same effort expended on culture of pancreas as in liver. Pahlman and Lieber et al. described neoplastic cells lines from exocrine pancreas; and Wallace and Hegre (1979) produced *epithelial monolayers* from fetal rate pancreas by a primary explant technique. These cultures remained free of fibroblasts for several days, and contained many endocrine cells.

The following protocol for the culture of pancreatic acinar cells has been contributed by Robert J. Hay.

Principles

Fractionated populations of pancreative acinar epithelia are caused to aggregate by gyration, the three-dimensional complexes are inoculated onto collagen-coated culture dishes, and two dimensional aggregated colonies are allowed to develop for subsequent study.

Outline

Dissociate guinea pig pancreatic tissue, filter the mixed cell suspension through nylon sieves, and layer the suspension over a BSA solution. After three sequential centrifugation/fractionation steps, collect the cell pellet and aggregate by gyration in a waterbath (2-24 hr). Inoculate onto collagencoated culture vessels, incubate, and observe.

Materials

Shaker-incubator, Model G-76 (New Brunswick)

Conical flasks, 25-ml (Bellco), siliconized.

Nylon mesh, 40 and 20-μm pore size (Tetco)

Dialysis tubing

HBSS

HBSS without Ca^{++} and Mg^{++} (HBSSDVC)

F-12K tissue-culture medium with 20% bovine calf serum (F12K-CS20)

Collagenase type 1 (Sigma); dissolve collagenase in BSSS-DVC (pH 7.2-7.4) to give 1,800 U/ml and dialyze overnight at 4°C against 20 × its volume with HBSS-DVC and a 12,000 MW exclusion membrane-filter, sterilize, and store in 10-20 ml aliquots at –70°C or below.

Trypsin 1:250 (ICN), 0.25% in citrate buffer, 3 g/l trisodium citrate 6 g/l NaCl, 5 g/l glucose, adjusted to pH 7.6 with NaOH

Bovine serum albumin, fraction V (Sigma)

Pipettes: 5-or 10-ml, wide bore (Bellco) collagen-coated culture dishes.

Protocol

1. Make up the dissociation fluid proper just before use by mixing 1 part of trypsin solution with 1 part of collagenase.

2. Aseptically remove the entire pancreas (0.4-1.0 gm) and place in HBSS, trim away mesenteric membranes and other extraneous matter, and mince into 1-3 mm^{-3} fragments.
3. Transfer to a siliconized, 25-ml conical flask in 5 ml of prewarmed, dissociation fluid and agitate at about 120 rpm for 15 min at 37°C in a shaker bath. Allow larger fragments to settle and decant the supernate over a sterile (autoclaved) nylon filter (40-μm pore size) fitted to a Buchner funnel. Repeat these dissociation steps two or three times with fresh fluid until most of the tissue has been dispersed.
4. Pool the cell suspensions and pass through a second filter/funnel combination using 20-μm nylon mesh in this case. It may be necessary to apply light suction by inserting the funnel into a vacuum filter flask during each filtration step.
5. Dispense the resulting fluid to 40-ml centrifuge tubes, take an aliquot for counting, and sediment the cells by centrifugation at 250 g for 5 min. Generally, 5-10 $\times$ 10^7 cells are obtained at this stage and 30 to 40% can be identified as acinar cells on the basis of zymogen droplet inclusions. These are readily visible by light and phase microscopy.
6. Resuspend the cell pellet in 10 ml HBSS-DVC and carefully layer 5 ml over each of two columns consisting of 35 ml of 4% BSA in HBSS-DVC in centrifuge tubes. Centrifuge at 100 $\times$ g for 5 min., discard the supernate, and repeat this fractionation step twice more, pooling and resuspending the pellets after each step.
7. Collect the cell pellets in 5 of 10 ml F-12K CS20. Yields of 0.9-3 $\times$ 10^7 viable cells (90%) are obtained, with 80 to 96% of the total being acinar.
8. Transfer the suspension to 25-ml siliconized conical flasks, 5 ml per flask; equilibrate with 5% CO_2 in air and place in a gyration-type water bath (shaker-incubator) at 37°C at 80 rpm for 2-24 hr.
9. After this period, allow the aggregates to settle for 3 to 5 min and discard the supernate. Resuspend the aggregates in F-12K-CS20 using a wide-bore pipette and take an aliquot for counting. Inoculation densities can be varied from 10^2-10^5 aggregates per cm^2 depending upon experimental design. Aggregates adhere and form colonies within 24-48 hr.

Variations

The methods has been applied for studies with guinea pig and human (transplant donor) tissues. Addition of human lung irradiated feeder-fibroblasts produces a marked stimulation (up to 500%) in ^{3}H-thymidine incorporation and prolongs survival at least by a factor of two.

Kidney

The separation of tubular epithelium from stroma has been one of the simpler problems in the field of epithelial cell culture. As these cells have *D-amino acid oxidase*, they will grow in D-valine while the stromal cells do not. Kidney tubular epithelium has also proved amenable to separation from stroma by *curtain electrophoresis* and *velocity sedimentation*. One of the simpler approaches is to treat the finely chopped kidney with collagenase overnight and then, following dispersal of the tissue by gentle pipetting, to allow the undisaggregated fragments of tubule and glomeruli to sediment through the more finely dispersed stroma. If the tubules are washed two or three times by repeating this differential sedimentation, a culture highly enriched for tubular epithelium and glomeruli may be obtained.

A continuous line of dog kidney epithelium has been described, and numerous monkey kidney primaries and cell lines are in regular use in vaccine laboratories. BSC-1 is an apparently normal epithelial line from African green monkey kidney, but human lines are rare. The success in growing kidney epithelium from many species may derive from the fact that they are of mesodermal origin: this may confer a different growth factor requirement from endodermally or ectodermally derived epithelium one more likely to be found in serum.

Bronchial and Tracheal Epithelium

There are a number of reports of primary culture of alveolar, bronchial, and tracheal epithelium including the use of floating collagen and pigskin, and Steele et al., (1978) were able to produce nontumorigenic continuous cell lines by treating tracheal epithelium with a phorbol ester. More recently Lechner and LaVeck (1985) have developed a low serum medium for clonal growth of normal lung and Carney et al., (1981) have developed a serum-free medium supplemented with hydrocortisone, insulin, transferrin, estrogen, and selenium (HITES medium) for small cell carcinoma of lung and with modifications for large cell and adenocarvinoma. These media are selective and do not support stromal cells.

The following protocol for the isolation and culture of normal human bronchial epithelial cells from autopsy tissue has been contributed by Moira A. LaVeck and John F. Lechner.

Principle

In the presence of serum, normal human bronchial epithelial (NHBE) cells cease to divide and, furthermore, terminally differentiate. The serum-free medium which has been optimized for NHBE cell growth does not support lung fibroblast cell replication, thus permitting the establishment of pure NHBE cell cultures.

Outline

This procedure involves first explanting fragments of large airway tissue in a serum-free medium (LHC-9) for initiating and subsequently propagating fibroblast-free outgrowths of NHBE cells; four subculturings and population doublings is routine.

Materials

Culture medium (sterile, filtered):

L-15 (Gibco)

LHN-9

HB (1985)

Bronchial tissue from autopsy of noncanerous donors

Plastic tissue culture dishes (60 and 100 mm, sterile)

Mixture of human fibronectin (Collaborative Research/Collagen (Vitrogen 100, Collagen Corp.)/ crystallized bovine serum albumin (Miles Biochemical) in LHC basal medium.

Gloves (human tissue can be contaminated with biologically hazardous agents)

Humidified CO_2 incubator at 36.5°C

Scalpel no. 1621 Becton Dickinson (sterile)

Surgical scissors (sterile)

Half-curved microdissecting forceps (sterile)

High O_2 as mixture (50% O_2, 45% N_2, 5% CO_2)

Controlled atmosphere chamber.

Rocket platform

0.02% Trypsin (Copper Biomedical)/0.02% EGTA (Sigma)/1% polyvinyl-pyrolidine (U.S. Biochemical Corp.) solution (sterile, filtration)

Pipettes (10 and 25 ml, sterile)

Phase contrast inverted microscope

Protocol

1. Coat culture dish with 1 ml of FN/V/BSA mixture per 60-mm dish and incubate in a humified CO_2 incubator at 36.5°C for at

least 2 hr (not to exceed 48 hr). Vacuum aspirate the mixture and fill the dish with 5 ml culture medium.

2. Aseptically dissected lung tissue from noncancerous donors autopsied within the previous 12 hr is placed into ice-cold L-15 medium for transport to the laboratory where the bronchus is further dissected from the peripheral lung tissues.
3. Before culturing, scratch a square centimeter area at one edge of the surface of the culture dishes using a scalpel blade.
4. Open the airways (submerged in L-15 medium) with surgical scissors and cut (slice, not saw) with a scalpel into two pieces, 20 × 30 mm.
5. Using a scooping motion to prevent damage to the epithelium, pick up the moist fragments and place epithelium-side-up onto the scratched area of the 60-mm dish. Remove medium and incubate at room temperature for 3 to 5 min to allow time for fragments to adhere to the scratched areas of the dishes.
6. Add 3 ml of HB medium to each dish and place in a controlled atmosphere chamber. Flush chamber with a high O_2 gas mixture and place on a rocker platform. Rock chamber at 10 cycles per minute, causing the medium to intermittently flow over the epithelial surface. Incubate rocking tissue fragments at 36.5°C, changing the medium and atmosphere after 1 d and 2-d intervals for 6-8 d. This step improves subsequent explant cultures by reversing ischemic damage to the epithelium that occurred form time of death of the donor until the tissue was placed in ice-cold L-15 medium.
7. Before explanting, scratch seven areas of the surface of each 100-mm culture dish with a scalpel. Coat the culture dish surface with the FN/V/BSA mixture and aspirate as before.
8. Cut the most ischemia-reversed fragments into 7 × 7 mm pieces and explant epithelium-side-up on the scratched areas. Incubate at room temperature without medium for 3 to 5 min as before.
9. Add 10 ml of LHC-9 medium to each dish and incubate explants at 36.5°C in a humidified 5% air/CO_2 incubator. Replace spend medium with fresh every 3 to 4 d.
10. After 8 to 11 d of incubation, when epithelial cell outgrowths radiate from the tissue explants more than 0.5 cm, transfer the explants to new scratched and FN/V/BSA coated culture dishes to produce new outgrowths of epithelial cells. This step can be repeated up to seven times with high yields of NHBE cells.

11. Incubate the postexplant outgrowth cultures in LHC-9 medium for an additional 2 to 4 d before trypsinizing (with trypsin/EGTA/PVP solution) for subculture or for experimental use.

Prostate

Propagation of prostatic cell was not reported before the introduction of serum-free selective media. There have been reports of primary culture systems, but these were not successfully subcultured.

The following protocol for the primary culture of rat prostate epithelial cells has been contributed by W.L. McKeehan, W. Alton Jones.

Principle

Isolated normal rat prostate epithelial cells are a valuable model to study the cell biology of maintenance, growth and function of normal prostate under investigator-controlled and defined conditions. Normal cells serve as the control cell type out of which prostate adenocarcinoma cells arise. Conditions similar to those described here have also been useful for primary culture of epithelial cells for transplantable rat tumors. Key features of this method are specific support of epithelial cells by improved nutrient medium and hormone-like growth factors, while concurrently inhibiting fibroblast outgrowth due to deficient or inhibitory properties of the medium.

Outline

Remove and prepare prostates for cell culture (30 min). Incubate with collagenase (1 hr) and collect single cells and small aggregates of cells (1 hr). Inoculate and culture cells to monolayer (7d).

Materials

10-12 week-old male rats

Shaking waterbath at 37°C

Centrifuge

Hemocytometer or Coulter counter

Sterile

60-mm petri dishes (glass or plastic)

Type I collagenase (Sigma)

MSS (media salt solution)

Penicillin

Kanamycin

Scissors

Syringe and 14-g cannula

25-ml Erlenmeyer flasks
1-mm wire mesh screen made to fit a
50-ml conical centrifuge tube
Fetal calf or horse serum
50-ml plastic conical centrifuge tubes nylon screen filters of 253, 150, 100 and 41 μm to fit 50-ml conical tubes.
Nutrient medium WAJC 404 McKeehan et al., 1984; Chaproniere and [McKeehan, 1986].
25-well plastic tissue culture dishes
Cholera toxin
Dexamethasone
Epidermal growth factor (EGF)
Ovine or rat prolactin
Insulin
Partially purified or purified
Prostatropin (prostate epithelial cell growth factor)

Protocol

1. Aseptically remove desired lobes of the prostate and place in a sterile 60-mm petri dish.
2. Trim fat from lobes and weigh.
3. Add 2 ml collagenase at 675 U/ml in MSS containing 100 μ/ml penicillin and 100 μg/ml kanamycin.
4. Mince with scissors to approximately 1-mm pieces, small enough to fit through a 14-g cannula.
5. Using a syringe and 14-g cannula, transfer minced tissue fragments to a sterile 25-ml Erlenmeyer flask. Add collagenase to 1 ml per 0.1 g original wet tissue weight.
6. Incubate for 1 hr at 37°C on a shaking water bath.
7. Aspirate the digested suspension three times through a 14-g cannula and then pass the suspension through a coarse (1-mm) wire mesh screen fitted to 50-ml plastic conical tubes to remove debris and undigested material. Rinse through with an equal volume of MSS containing 5% whole fetal calf serum or horse serum.
8. Collect cells by centrifugation at 100 g for 5 min at 4°C.
9. Resuspend pellet in 5 ml of MSS plus 5% serum and pass the suspension successively through nylon screen filters of mesh sizes 253, 150, 100, and 41 mm. Wash each screen with 5 ml of MSS plus 5% serum.

10. Collect cells by centrifugation and resuspend in 5 ml of nutrient medium WAJC 404.
11. Count cells and adjust concentration to 4 × 10^6 cells/ml.
12. Innoculate 50 ml containing 2 × 10^5 cells into each well of a 24-well plate (area = 2 cm^2) containing 1 ml medium WAJC 404 and 10 ng/ml cholera toxin, 1 μM dexamethasone, 10 ng/ml EGF, 1 μg/ml prolactin, 5 mg/ ml insulin, 10 mg/ml prostatropin, 100 U/ml penicillin, and 100 Mg/ml streptomycin. Partially purified sources of prostatropin can be substituted as described in McKeehan et al., (1984) and Chaproniere and McKeehan (1986).
13. Incubate in a humidified at 95% air and 5% CO_2 at 37°C. Change medium at days 3 and 5. Cells should be near confluent by day 7.

Variations

This procedure can be applied to culture of human prostate epithelial cells with modifications described in Chaproniere and McKeehan (1986).

Mesenchymal Cells

I will include here those cells which are derived from the embryonic mesoderm, but exclude the hemopoietic system, which will be discussed below. This group includes the structural and vascular cells.

Connective Tissue

These cells are generally regarded as the weeds of the tissue culturist's garden. They survive most mechanical and enzymatic explanation techniques and may be cultured in many of the simplest media, such as Eagle's basal medium.

Although cells loosely called fibroblasts have been isolated from many different tissues and assumed to be connective tissue cells, the precise identity of cells in this class remains somewhat obscure. Fibroblast lines, e.g, 3T3 from mouse, produce types I and III collagen and release it into the medium. While collagen production is not restricted to fibroblasts, synthesis of type I in relatively large amounts is characteristic of connective tissue. However, 3T3 cells can be induced to differentiate into adipose cells. It is possible that cells may transfer from one lineage to another under certain conditions, but such trans differentiation has rarely been confirmed. It is more likely that mouse embryo fibroblastic cell lines are primitive mesodermal cells which may be induced to differentiate in more than one direction.

Human, hamster, and chick fibroblasts are morphologically distinct from mouse fibroblasts as they assume a spindle-shaped morphology at confluence, producing characteristic parallel assays of cells distinct from the pavement-like appearance of mouse fibroblasts. The spindle shaped cell may represent a more highly committed precursor and may be more correctly termed a fibroblast, NIH3T3 cells may become spindle shaped if allowed to remain at high cell density.

It has also been suggested that fibroblastic cell lines may be derived from vascular pericytes, connective tissue-like cells in the blood vessels, but in the absence of the appropriate markers, this is difficult to confirm.

It is clearly possible to cultivate cell lines, loosely termed fibroblastic, from embryonic and adult tissues, but these should not be regarded as identical or classed as fibroblasts without confirmation with the appropriate markers. Collagen, type I, is one such marker. Thy I antigen has also been used, although this may also appear on some hemopoietic cells.

Adipose Tissue

Although it may be difficult to prepare cultures from mature fat cells, differentiation may be induced in cultures of mesenchymal cells by maintenance of the cells at a high density for several days. An adipogenic factor in serum appears to be responsible for the induction.

Muscle

Myoblasts from the three main categories of muscle may be grown in culture. Skeletal and cardiac myoblasts may be prepared from chick embryo Yaffe (1986) and other have described the stages of differentiation in these cultures.

Cardiac myoblast cells will also progress through differentiation *in vitro* and can be seen to contract rhythmically a few days after explanation from the embryo, although they tend to lose this capacity with continued subculture. Polinger (1970) used the differential rate of attachment to reduce fibroblastic contamination of primary cultures.

Smooth muscle cells may be cultured from blood vessels following disaggregation in *trypsin* or *collagenase*. Gospodarowicz et al. (1976) described cloned cell lines from bovine aorta derived from scraped or collagenase-treated tissue. Yasin et al. (1981) obtained cell lines from adult skeletal muscle and showed that they retained specialized markers.

Muscle cells may be identified by a number of antigenic markers including myosin and tropomyocin. Actin is not a good marker as it can be found in most cells. *Creatine phosphokinase* activity increases as muscle cells differentiate. The most obvious property of all is spontaneous contraction, which is observed in both skeletal and cardiac muscle.

In common with other cells with excitable membranes (and some hemopoietic cells), muscle cells may be stained selectively with the fluorescent dye merocyanine 540.

The following protocol for the culture of skeletal muscle has been contributed by Jane plumb.

Principle

Skeletal muscle cultures can be established from either embryonic or newborn animal tissues. The muscle is dissected free of skin and bone and disaggregated by enzymatic digestion. Myoblasts are plated out at a high density to minimize fibroblast growth and fusion of the myotubes occurs within 5 to 6 days.

Outline

Remove muscle tissue from animals, mince, and incubate with collagenase for 24 hr. Disperse, remove collagenase, add medium, and seed cultures.

Materials

Scissors
Forceps
Scalpels
Universal containers or centrifuge tubes
Petri dishes
25-cm^2 culture flasks
Primaria petri dishes, 50 mm (Falcon)
Plastic box
Centrifuge
PBSA + antibiotics
Serum-free medium (Ham's F10: DMEM, 50:50)
Culture medium: as above plus 10% fetal bovine serum
Collagenase (2,000 U/ml) Worthington, CLS, Sigma, 1A
3-4 day-old rats
Pipettes

Protocol

1. Kill rats by cervical dislocation and wash thoroughly with 70% alcohol. Transfer to a sterile paper towel.
2. Use sterile instruments to remove skin. Start with a transverse cut around the body in the abdominal region. Then peel skin back over the fore and hind limbs.

3. Cut off fore and hind feet and remove fore and hind limbs at shoulder and hip joints respectively. Place limbs in PSA + antibiotics in a sterile container and keep on ice. Repeat procedure for up to 4 rats at a time.
4. Transfer container to a laminar flow hood. Remove PBSA and place limbs in a fresh sterile container. Wash four times with 20 ml PBSA and antibiotics.
5. Place the limbs in a sterile petri dish and add 5 ml of serum-free medium. Place 5 ml of serum-free medium in a second petri dish.
6. Remove fat from limbs and dissect muscle tissue away from the bone. Transfer tissue to the second petri dish.
7. Mince muscle tissue with crossed scalpels into pieces about 1 mm^3 and transfer medium and tissue to a sterile container.
8. Allow tissue to settle for about 5 min and then remove medium. Add 10 ml of serum-free medium and pipette medium plus tissue into a 25-cm^2 culture flask. Add collagenase (2,000 U) and incubate for 24 hr at 37°C.
9. Pipette vigorously to disperse tissue and transfer suspension to a sterile universal container or centrifuge tube. Centrifuge for 5 min at 500 g.
10. Remove supernatant and resuspend pellet in culture medium containing fetal calf serum (10%) and antibiotics (not streptomycin since this may inhibit myoblast fusion). Seed cultures at a high density in Primaria petri dishes. Tissue from four rats is sufficient for 30 × 50-mm dishes.
11. Put dishes in a plastic box and place in a CO_2 incubator at 36.5°C. Feed daily.

Characterization of cultures

Fusion of the myoblasts occurs after 4 to 5 d and can be observed with the aid of a phase contrast microscope. Provided that the cells are plated at a high density, fibroblast contamination is less than 20% of the total cell population. Fusion can be quantified by monitoring the production of the myosin heavy chain.

Variations

Cultures of skeletal muscle from chick embryos can be established in a similar manner. Fibroblast contamination can be reduced further by preplating the culture for 1 hr in a standard tissue culture flask prior to culture in Primaria petri dishes. This procedure reduces the overall yield of muscle cells. Human muscle cultures can be derived from primary explant cultures prepared from muscle biopsies of both normal and diseased tissues.

Cartilage

Coon and Cahn (1966) described a technique for the cultivation of cartilage-synthesizing cells from chick embryo somites. Cahn and Lasher (1967) later used this system for analysis of the involvement of DNA synthesis as a prerequisite for cartilage differentiation. Chondrocytes respond to stimulation of growth by both EGF and FGF but ultimately lose their differentiated function.

The following protocol for the culture of human chondrocyte cultures has been contributed by Edith Schwartz.

Principle

Chondrocytes are embedded in a dense matrix of proteoglycans which must be digested by sequential enzymatic treatment to release the relatively small cellular compartment of the tissue. The cells are then cultured in slightly alkaline Ham's medium with serum and increased Mg^{2+}.

Outline

Finely chopped fragments of cartilage are treated twice with hyaluronidase, trypsin, and collagenase to remove the matrix and make them available to more prolonged collagenase digestion. The cells which are then released are propagated by conventional monolayer techniques in slightly alkaline Ham's F12 medium with an elevated Mg^{2+} concentration.

Materials

Gey's balanced salt solution (GBSS) pH 7.0; if contamination becomes a problem, add antibiotics (100 U/ml penicillin, 100 μg/m streptomycin or 50 μg/ml gentamycin, with or without 100 μg/ml mycostatin) to digestion mixture.

Ham's F12 with 12% fetal bovine serum, 2.3 mM Mg^{2+}, 100 U/ml penicillin, 100 μg/ml streptomycin SO_4, pH 7.6

Serum-free medium: as above, without serum

0.5% testicular hyaluronidase in GBSS with 100 U/ml penicillin and 100 μg/ml streptomycin

0.2% collagenase in GBSS

0.2% trypsin (Sigma type IX) (for tissue digestion) in GBSS

0.25% trypsin (Sigma type XI) in Ca^{2+} - and Mg^{2+} -free Tyrode's (to be used for cell passage only)

Protocol

1. Transfer cartilage to a 50-cm sterile glass petri dish.
2. Cover cartilage with GBSS.
3. With the use of two scalpels (one no. 20 blade and one no. 1! blade), cut cartilage into 1-2 mm³ segments.
4. Transfer cut segments to a second glass petri dish. Cover with GBSS.
5. When all segments have been cut and combined, remove GBSS and cover tissue with 4 ml of hyaluronidase for 5 min at room temperature.
6. Remove the hyaluronidase solution and replace with 8 ml fresh hyaluronidase solution for an additional 10 min at room temperature.
7. Remove the hyaluronidase solution at the end of this time and wash the tissue fragments two times with 5 ml of GBSS.
8. At the end of this procedure, use GBSS to transfer the cartilage piece to a 25-ml Corex screw-top tube.
9. Remove the GBSS and wash the tissue with 2 ml of trypsin solution. Discard wash.
10. Wash the tissue again with an additional 2 ml of trypsin solution and discard wash.
12. Incubate the tissue with 4 ml of trypsin at 37°C for 30 min with stirring.
13. Wash the tissue fragments for 5 min with 2 ml of 0.2% collagenase dissolved in GBSS. Discard collagenase wash.
14. Incubate the tissue with 4 ml of collagenase for 30 min at 37°C.
15. Remove and discard the supernatant.
16. Add an additional 4 ml of collagenase to the tissue sample and incubat at 37°C for 90 min.
17. Remove the supernatant and centrifuge at 600 g for 8 min to pellet the cells.
18. Resuspend the cells in 8 ml Ham's F12 medium with 20% FCS and centrifuge at 180g for 1 min to sediment undigested matrix particles.
19. Remove the supernatant to another tube and centrifuge it at 600 g for 10 min to sediment the cells.
20. Resuspend the cells in 8 ml of F12 medium with 12% FBS and inoculate a 75-cm² flask.
21. Add an additional 4 ml of F12 medium with 12% FBS to bring the final volume to 12 ml.

22. Repeat steps 16-21 two or three more times as warranted by the amount of tissue in the sample.

Bone

Although bone is mechanically difficult to handle, thin slices treated with EDTA and digested in collagenase give rise to cultures of osteoblasts which have some functional characteristics of the tissue. Antiserum against collagen has been used to prevent fibroblastic overgrowth without inhibiting the osteoblasts. Propagated lines have been obtained from osteosarcoma but not from normal osteoblasts.

The following protocol for the culture of bone cells has been also contributed by Edith Schwartz.

Principle

Bone cultures suffers from the inherent problem that the hard nature of the tissue makes manipulation difficult. However, conventional primary explant culture or digestion in collagenase and trypsin releases cells which may be passaged in the usual way.

Materials

Ham's F12 medium with 12% fetal bovine serum, 2.3 mM Mg^{2+}, 100 U/ml penicillin, and 100 Mg/ml streptomycin SO_4

Trypsin solution to be used for cell passage: dissolve 125 mg trypsin (sigma type XI) in 50 ml of Ca^{2+}- and Mg^{2+}-free Tyrode's solution. Adjust to pH 7 and filter sterilize. Place aliquots of 5 and 10 ml in Pyrex tubes and store at −20°C

Digestion solution for the isolation of osteoblasts:

To prepare solution A, dissolve 8.0 g NaCl, 0.2 g KCl, 0.05 g NaH_2PO_4 H_2O in 100 ml distilled water

To prepare (Collagenase-trypsin solution) solution B, dissolve 137 mg Collagenase (type I, Worthington Biochemicals), and 50 mg trypsin (Sigma, type III) in 10 ml solution A. Adjust to pH 7.2 and then bring to 100 ml with distilled water. Filter sterilize and distribute into 10-ml aliquots which are stored at −20°C

Scalpels

Forceps

9-cm petri dishes

25- or 75-cm² flasks (Corning, Falcon, Nunc)

Outline

Small fragments of tissue are allowed to adhere to the culture flask by incubation in a minimal amount of medium. The adherent explants are then flooded and the outgrowth monitored.

Protocol

1. Obtain bone specimens from the operating room.
2. Rinse the tissue several times at room temperature with sterile saline.
3. If the bone cannot be used immediately, cover the bone specimen with sterile Ham's F12 medium containing 12% fetal calf serum and penicillin and streptomycin sulfate. The bone may be stored overnight at 4°C.
4. The next morning, rinse the tissue with Tryode's solution containing penicillin (100 U/ml) and streptomycin sulfate (100 μg/ml).
5. Place the one in a petri dish and with the use of scalpel and forceps, remove trabeculae and place in a second petri dish.
6. Add 10 ml Tyrode's solution over the excised trabeculae. Rinse the trabeculae several times with Tyrode's solution until blood and fat cells are removed.
7. To initiate explant cultures, prepare a 25-cm^2 flask by pre-incubating with 2 ml of complete medium for 20 min to equilibrate the medium with the gas phase. Adjust the pH as necessary with CO_2 or 4.5% $NaHCO_3$, or HCl.
8. Cut the trabeculae into fragments of 1-3 mm^3.
9. Remove the preincubation medium from the flask and add 2.5 ml of fresh Ham's F12 medium to the flask. Transfer between 25 and 40 fragments of trabeculae to the flask.
10. With the flask in an upright position, slide the explant pieces with the aid of an inoculating loop along the base of the flask and distribute the explant pieces evenly.
11. Permit the flask to remain upright for 15 min at 37°C.
12. Slowly restore the flask into a normal horizontal position. The explant pieces will stick to the bottom of the flask.
13. Leave the flask in the horizontal position at 37°C for 5 to 7 d. After this period, check for outgrowth and replace the medium of the flask with fresh medium.

    To avoid detachment of the explant pieces, lift the flask slowly into the vertical position before carrying from the incubator to the hood or to the microscope.
14. To maintain cultures, change medium two times per week.
15. When confluency is reached, the explant pieces are removed, the cell layer is trypsinized, and the cells are isolated by centrifugation and seeded into flasks or wells.

Monolayer Cultures from Disaggregated Cells

Outline

Trabecular bone is dissected down to 2-5 mm^3 and digested in collagenase and trypsin. Suspended cells are seeded into flasks in F12 medium.

Protocol

1. Wash trabecular bone specimens repeatedly with Tyrode's solution to remove the fat and blood cells. The trabeculae are excised with scalpel and forceps under sterile conditions.
2. After collecting as much bone as possible, wash the remaining blood and fat cells away by rinsing three times with Tyrode's solution.
3. Wash the cut trabeculae with F12 medium containing fetal calf serum.
4. Place the bone pieces in a small sterile bottle with a magnetic stirrer and add 4 ml of digestion solution (this should cover the bone specimens).
5. Stir the solution containing bone fragments at room temperature for 45 min.
6. Remove the suspension of released cells and discard since these cells are most likely to contain fibroblasts.
7. Add a second aliquot of 4 ml of digestion solution to the bone fragments and stir the mixture at room temperature for 30 min.
8. Collect the digestion solution from bone fragments and centrifuge for 2 min at 580 g at room temperature.
9. After removing the supernatant, suspend the cells in 4 ml of Ham's F12 medium with 20% fetal calf serum and count the cells.
10. Centrifuge at 580 g for min and resuspend the cells in 4 ml of complete medium. This will become the inoculum.
11. Preincubate 75-cm^2 flask for 20 min with 8 ml of complete F12 medium to equilibrate with the gas phase.
12. Remove the preincubation medium and add 2 ml of complete F12 medium.
13. Add 4 ml of medium containing the cell suspension. The inoculum should contain 6-10,000 cells per cm^2 of surface area.
14. Finally, add an additional 6 ml of Ham's F12 medium to given a total volume of 12 ml.

16. In the interim, add an additional 4 ml of digestion solution to the remaining bone pieces and repeat the digestion for 30 min. The

released cells are harvested and, if necessary, the digestion step is repeated several more times. With large amounts of bone the digestion period can be increased to 1-3 hr. Cell counts are performed after each digestion period and the released cells are used to inoculate a different flask.

Passage of cells in culture

1. Remove the explant pieces
2. Remove the medium and rise the cell layer with PBS, 0.2 ml per cm^2.
3. Add trypsin to the flsk, 0.1 ml/cm^2, and incubate at 37°C until the cells have detached and separated from one another. This is monitored under the microscope. In general, a 10-min incubation is sufficient.
4. Transfer the released cells to a centrifuge tube with an equal volume of Ham's F12 medium with 20% fetal calf serum (FCS).
5. Centrifuge at 600 g for min.
6. Discard the supernatant and resuspend the cells in complete medium by gentle repeated pipetting.
7. Set one aliquot aside for the determination of cell concentration and another for DNA determination.
8. Inoculate the remaining cells into culture flasks or wells which have previously been equilibrated with medium. The cells should reattach within 24 hr.

Endothelium

Endothelium has been successfully cultured by collagenase perfusion of bovine aorta and human umbilical vein trypsinization of white matter from rat cerebral cortex, and microdissection of adrenal cortex. In the author's laboratory, an endothelial cell line has been developed from a human anaplastic astrocytoma by collagenase digestion. The astrocynoma cell were overgrown during serial passage.

Endothelium can be characterized by the presence of factor VIII antigen, type IV collagen. Weibel-Palade bodies and, sometimes, the formation of tight junctions, although the last feature is not always demonstrated readily in culture.

Endothelial cultures are good models for contact inhibition and density limitation of growth as cell proliferation is strongly inhibited after confluence is reached.

Much interest has been generated in endothelial cell culture because of the potential involvement of endothelial cells in vascular disease, blood vessel repairs, and angiogenesis in cancer. Folkman and Haudenschild (1980) described the development of three-dimensional structures resembling capillary blood vessels derived from pure endothelial lines *in vitro*. Growth factors, including angiogenesis factor derived from Walker 256 cells *in vitro*, play an important part in maintaining proliferation and survival, so that secondary structures can be formed.

The following protocol for the culture of large vessel endothelial cells has been contributed by Bruce Zetter.

Principle

Endothelial cells comprise a single cell layer at the inner surface of blood vessels. The vessels most commonly used to obtain cultured endothelial cells are the bovine aorta, bovine adrenal capillaries, rat brain capillaries, human umbilical veins, and human dermal and adipose capillaries. Although all endothelia share some properties, significant differences exist between the endothelial cells of large and small blood vessels.

Outline

Endothelial cells released by collagenase incubated within the blood vessel are cultured on a gelatin-coated substrate in medium supplemented with mitogens (human and bovine capillary) or without mitogens (bovine large vessel).

Materials

Aseptically isolated blood vessels, preferably in 10-cm sections, approximately 5-mm diameter. If asepsis cannot be guaranteed, clamp both ends and dip in 70% alcohol for 30 sec

Collagenase: 0.25% crude collagenase, approximately 200 U/mg, in PBSA containing 0.5% bovine serum albumin

Dulbecco's modified Eagle's medium supplemented with 10% calf serum and antibiotics.

10-cm tissue culture grade petri dishes or 75-cm^2 flasks coated with 1.5% gelatin in PBSA: incubate overnight in gelatin solution, remove gelatin, add medium with serum, and incubate until cells are ready

Two clamps or hemostats, 25 mm

Sharp scissors, 50 mm

Protocol: Isolation of endothelial cells

1. Ligature one end of a 10-cm section of blood vessel 2-10 mm diameter to a 5-ml plastic syringe.

2. Wash vessel with 20 ml PBSA to remove blood.
3. Run in collagenase solution until it appears at bottom end, clamp lower end with hemostat, and incubate at room temperature for 10 min.
4. Cut vessel above clamp with sharp scissors and collect the collagenase in a 10-cm Petri dish.
5. Rinse lumen of vessel with 10 ml PBSA and add to collagenase.
6. Centrifuge pooled digest and wash at 100 g for 5 min.
7. Wash twice by resuspending in medium and centrifuging.
8. Resuspend final pellet in growth medium and seed into gelatin coated dishes or flasks, approximately one 10-cm section of blood vessel, 5-mm diameter, per 75-cm flask or 10-cm diameter dish.
9. Subculture by conventional trypsinization.

Identification

This is proved by production of factor VIII, angiogensin converting enzyme, uptake of acetylated low-density lipyprotein, presence of Weibel-Palade bodies and expression of endothelial specific cell surface antigens.

Variations

Human endothelial cells are cultured in medium 199 with 10-20% human serum, 25 μg/ml hypothalamus derived endothelial mitogen and 90 μg/ml heparin. Capillary endothelial cells also require direct addition of an endothelial mitogen or conditioned medium from a tumor cultures.

NEURECTODERMAL CELLS

Neurones

Nerve cells appear more fastidious in their choice of substrate than most other cells. They will not survive well on untreated glass or plastic but will demonstrate neurite outgrowth in collagen and poly-D-lysine. Neurite outgrowth is encouraged by a polypeptide nerve growth factor (NGF) and a factor secreted by glial cells immunologically distinct from NGF.

Cell proliferation has not been found in cultures of neurons even with cells from embryonic stages where mitosis was apparent *in vivo*. Much of the work on nerve cell differentiation has, therefore, been performed on neuroblastoma cell lines or in glial-neuronal hybrids. This remains an intriguing area with many unsolved problems.

The following protocol for the monolayer culture of cerebellar neurons has been contributed by Bernt Engelsen and Rolf Bjerkvig.

Principle

Cerebellar granule cells in culture provide a well-characterized cell population suited for morphological and biochemical studies. The cells are obtained from the cerebella of 7 or 8-day-old rats, and initial growth inhibition of nonneuronal cells is obtained by a short addition of cytosine arabinoside to the cultures.

Outline

The cerebella from four to eight neonates are cut into small cubes and trypsinized for 15 min at 37°C in Hanks' balanced salt solution. The cell suspension is seeded in poly-L-lysine coated culture wells/flasks.

Materials

Dulbecco's modification of Eagle's medium containing

10% heat inactivated fetal calf serum 30 mM glucose

L-glutamine 293.2 mg/l

24.5 mM HCl

100 mU/l insulin (Sigma I-1882) 7 μM p-aminobenzic acid (Sigma A-3659)

100 μg/ml gentamycin

Hanks' balanced salt solution with 3g/l BSA (HBSS)

Poly-L-lysine, mol wt > 300,000 (Sigma P-1524)

Cytosine arabinoside (Cytostar, upjohn; powder)

Trypsin (type II) (0.025% in HBSS)

Silicone (Aquasil, Pierce 42799)

35-mm tissue culture petri dishes

Scalpels, scissors, and surgical tweezers (sterile)

Waterbath

Pasteur pipettes and 10-ml pipettes (sterile)

12- and 50-ml sterile test tubes

Protocol

1. *Siliconization of Pasteur pipettes*: dilute the Aquasil solution in distilled deionized water to a 0.1-1% concentration. Dip the pipettes into the solution or flush on the inside. Air dry for 24 hr, or for several min at 100°C, and sterilize.
2. *Poly-L-lysine treatment of culture dishes*: dissolve the poly-L-lysine in distilled water (10 mg/l) and sterilize by filtration. Add 1 ml of poly-L-lysine solution to each of the 35-mm petri dishes. Remove the poly-L-lysine solution after 10-15 min and add 1-15 ml of culture medium. Place the culture dishes in the incubator (minimum 2 hr) until seeding the cells.

3. *Preparation of cells :*
   (a) Dissect out the cerebella aseptically and place them in HBSS. Mince the tissue with scalpels into small cubes approximately 0.5 mm^3.
   (b) Transfer to test tubes (12 ml) and wash three times in HBSS. Allow the tissue to settle to the bottom of the tubes between each washing.
   (c) Add 10 ml of 0.025% trypsin to the tissue and incubate in a waterbath for 15 min at 37°C.
   (d) Transfer the trypsinized tissue to a 50-ml test tube and add 20 ml of growth medium to stop trypsin action.
   (e) Shear tissue by trituration through a siliconized Pasteur pipette, until a single cell suspension is obtained.
   (f) Let the cell suspension stay in the test tube for 3-5 min allowing small clumps of tissue to settle to the bottom of the tube. Remove these clumps with a Pasteur pipette.
   (g) Centrifuge the single cell suspension at 200 g for 5 min and aspirate off the supernatant.
   (h) Resuspend the pellet in growth medium and seed the cells at a concentration of 2.5-3.0 × 10^6 cells/dish.
   (i) After 2-4 d (best results usually after 2 4), incubate the cultures with 5-10 μM cytosine arabinoside for 24 hr. Then change to ordinary culture medium.

Analysis

Neurons can be identified by immunological characterization using neuron specific enolase antibodies or by using tetanus toxin as a neuronal marker. Astrocyte contamination can be quantified by using glial fibrillary acidic protein protein as a marker.

Variations

A single cell suspension can be obtained by mechanical sieving through nylon meshes of decreasing diameter, or by sequential trypsinization (i.e. 3 × 5-min trypsin treatment). Instead of HBSS, Puck's solution, Krebs, or other buffers with glucose can be used.

The following Protocol for Aggregating Cultures of Brain Cells has been Contributed by Rolf Bjerkvig

Principle

Aggregating cultures of fetal brain cells have been extensively used to study neural cell differentiation. The aggregating cells follow the same development sequence as observed *in vivo*, leading to an organized structure consisting of mature neurons, astyrocytes, and oligodendrocytes. A prominent neuropil is also formed. In tumor biology, the aggregates can be used to study brain tumor cell invasion *in vitro*.

Outline

Brains from fetal rats at day 17-18 of gestation are removed and prepared as a single cell suspension. Brain aggregates are formed by overlay cultures in agar coated multiwells. The cells in the aggregates will form a mature organoid brain structure during a 20-day culture period.

Materials

Dulbecco's modification of Eagle's medium containing
10% heat inactivated newborn calf serum
Four times the prescribed concentration of non-essential amino acids
L-glutamine 293.2 mg/l
Penicillin 100 μ/ml streptomycin 100 μg/ml
Phosphate buffered saline (PBS) with Ca^{2+} and Mg^{2+}.
Trypsin type II (0.025% in PBSA)
Waterbath
Agar (Difco)
Multiwell tissue culture dishes.
10-cm sterile petri dishes
12-ml sterile test tubes
Pasteur pipettes and 5-ml pipettes (autoclaved)
Scalpels, scissors, and surgical tweezers (sterile)
2 sterile 100-ml Erlenmeyer flasks

Protocol

1. *Medium-agar coating of microwells*: Prepare a 3% stock solution (3 g agar in 100 ml PBSA) in an Erlenmeyer flask. Heat the flask in boiling water until the agar is dissolved. Place an empty Erlenmeyer flask in boiling water and add 10 ml of hot agar solution. Then slowly add complete growth medium to the flask until a medium-agar concentration of 0.75% is reached. Add 0.5 ml of hot medium-agar solution to each well in the multiwell dish. Wait until the agar has cooled. The multiwell dishes can be stored in a refrigerator for 1 week.

2. *Preparation of Cells*:

(a) Dissect out aseptically the whole brains from a litter of fetal rats at day 17-18 gestation. Place the tissue in a 10-cm petri dish containing PBSA. Mince the tissue with scalpels into small cubes 0.5 cm^3.

(b) Transfer the tissue to a test tube and wash three times in PBSA. Allow the tissue to settle to the bottom of the tube between each washing.

(c) Add 5 ml of trypsin solution and incubate in a waterbath for 5 min at 37°C.

(d) Shear tissue by trituration through a Pasteur pipette approximately 20 times. Allow to settle for 3 min and transfer the clump-free milky cell suspension to a test tube containing 5 ml of growth medium. Five ml of new trypsin is added to the undissociated tissue and the trypsinization and dissociation procedure is repeated two more times.

(e) Spin the cell suspensions at 200 g for 5 min. Aspirate supernatant, resuspend, and pool cells in 10 ml of growth medium. Count the cells.

(f) Add 3 × 10^6 cells to each agar coated well. The volume of the overlay suspension should be 1 ml. Place the multidish in a CO_2 incubator for 48 hr.

(g) Remove the aggregates to a sterile 10-cm petri dish. Add 10 ml growth medium. Transfer larger aggregates individually to new agar coated multiwells by using a Pasteur pipette.

(h) Change medium every third day by carefully removing and adding new overlay medium. During 20 days in culture, the aggregates will become spherical and develop into an organoid structures.

Analysis

The next step is fixation and embedding in paraffin or epon for histological or electron microscopic evaluation. Oligodendrocytes, astrocytes, and neurons are identified by transmission electron microscopy or by immunohistochemical localization of myelin basic protein, glial fibrillary acidic protein, and neuron specific enolase, respectively.

Variations

A single cell suspension can be obtained by mechanical sieving through steel or nylon meshes. Reaggregation cultures can also be obtained using a gyratory shaker. A speed (about 70 rpm) is selected such that the cells are brought into vortex, thereby greatly increasing the number of collisions between cells. This movement also prevents cell attachment to the culture flasks.

GLIA

Greater success has been obtained in culturing *glial cells* from avian, rodent, and human brain. Embryonic and adult brain give cultures by trypsinization, collagenase digestion, and primary explant which closely resemble glia. Astrocytes, markers can be demonstrated for several subcultures although there is only one report that cell lines from human adult normal brain lines express the most specific marker, glial fibrillary acidic protein (GFAP). It is our experience that while some glial properties remain (high-affinity-γ-aminobutyric acid and glutamate uptake, *glutamine synthetase* activity), GFAP is lost. Oligodendrocytes do not readily survive subculture, but Schwann cells from optic nerve have been subcultured using cholera toxin as a mitogen, 1978.

Cultures of human glioma can also be prepared by mechanical disaggregation, trypsinization, or collagenase digestion. The right temporal lobe from human males appears to be marginally better than other regions of the brain, but most give a good chance of success. The glia/glioma system provides a good model for comparing normal and neoplastic cells under the same conditions.

There is good evidence that the cell lineage is the same, particularly between embryonic normal cells and tumor cells, but the position of cells within the lineage, as with many cultured cells, is still debatable.

The primary explant technique and collagenase digestion have both been found suitable for preparing glial cultures from normal fetal and adult brain. Satisfactory cultures may also be obtained from fetal and newborn rodent brain by the cold trypsinization techniques.

A number of gliomas have been cultured from rodents among with the C^6 deserves special mention. This cell line expresses the astrocytic marker, glial fibrillary acidic protein, in up to 98% of cells but still carries the enzymes glycerol phosphate dehydrogenase and 2'3' cyclic nucleotide phosphorylase, both of which are oligodendrocyteic markers. This appears to be an interesting example of a precursor cell tumor which can mature along two distinct phenotype routes simultaneously.

Linser and Moscona (1980) separated the Muller cells of the neural from pigmented retina and neurons and demonstrated that full functional development could not be achieved unless the Muller cells (astroglia) were recombined with neurons from the retina. Neurons from other regions of the brain were ineffective.

Endocrine Cells

The problems of culturing endocrine cells are similar to the culture of any other specialized cell but accentuated because the relative number of secretory cells may be quite small. Sato and colleagues cultured functional adrenal and pituitary cells from rat tumors by mechanical disaggregation of the tumor and regular monolayer culture. The functional integrity of the cells was retained by intermittent passage of the cells as tumors in rats. These lines are now fully adapted to culture and can be maintained without animal passage and in some cases in fully defined media.

Fibroblasts have been reduced in cultures if pancreatic islet cells by treatment with ethylmercurithiosalicylate and have also been purified by density gradient centrifugation. These cells apparently produce insulin but not as propagated cell lines.

Pituitary cells, which continue to produce pituitary hormones for several subcultures, have been isolated from the mouse but in our experience normal human pituitary cells do not survive well and even pituitary adenoma cells gradually lose the capacity for hormone synthesis.

Melanocytes

Pigment cells were cultured successfully by Coon from chick pigmented retina and propagated over many generations. As with the chick embryo cartilage cells, a fraction derived from embryo extract was required for the function differentiation of these cells.

Until recently, other normal pigment cells have proven difficult to culture although cultures were obtained from human uveal *melanocytes*. Pigment cells from skin do not survive readily without the appropriate growth factors although cultures can be obtained from melanomas with a reasonable degree of success is obtained with secondary growth from lymph nodes, or from distant metastatic recurrences. Sato (1979) has described conditions for serum-free culture of cell lines from human and murine melanoma.

The following protocol for the culture of human melanocytes has been contributed by Barbara A. Gilchrest.

Principle

The greater substrate dependency of cultured keratinocytes has been utilized to obtain preferential melanocyte attachment and growth in a hormone-supplemented medium containing a potent, previously undescribed *melanocyte growth factor* (MGF) extracted from bovine hypothalamus. The system has been modified to obviate the problem of keratinocyte contamination while supporting good melanocyte proliferation and pigment production in the absence of tumor promoters or chemotherapeutic agents with minimal serum supplementation. With conventional serum supplementation (5-20%), melanocyte growth is far better than reported in other systems.

Outline

Epidermis, stripped from small fragments of skin following cold trypsinization, is dissociated in EDTA and cultured in serum-free medium supplemented with MGF.

Materials

Humidified incubator (37°C, 8% CO2)
Waterbath
Coulter counter or hemocytometer
Tissue culture dishes (Falcon, Becton, Dickindon), 100-, 60-, and 35-mm diameter sterile pipettes
15- and 50-ml centrifuge tubes
PBSA
Fetal bovine serum
Hormone-supplemented medium: medium 199
EGF (Collaborative Research), 0.1ml, 10 μg/ml stock concentration, 10 ng/ml final concentration
Transferrin (Sigma), 0.1 ml, 10 Mg/ml
Stock, 10 μg/ml final
Insulin (Sigma), 0.1 ml, 10 mg/ml
Stock, 10 μg/ml final
Triiodothyronine 0.1 ml, 10–6M stock, 10^{-9} M final
Hydrocortisone 0.5 ml, 200 × stock. 1.4 × 10^{-6} M final
Cholera toxin (Calbiochem), 1.0 ml, 10–8M stock, 10^{-9}M final
MFG 5.0 ml, 2 mg/ml stock, 100 μg/ml final.

Preparation of melanocyte growth factor (MGF)

1. Wash bovine hypothalami in dH_2O (21 kg/tissue) until grossly free of blood.
2. Homogenize in Waring blender for 3 min in 0.15 M Cl (11/kg tissue)

3. Extract homogenate for 2 hr at 4°C in beaker with magnetic stirrer bar at moderate speed.
4. Centrifuge at 10,000 g for 30 min at 4°C in in 500-ml aliquots.
5. Decant supernatant (containing MGF) and delipidate with streptomycin sulphate. Prepare streptomycin sulphate 7.5/kg tissue in 50 ml chilled dH_2O, pH chilled 8.0, and add to extract while stirring over a 5-min period. Maintain extraction preparation. At pH 6.5-7.5 at all times.
6. Centrifuge at 20,000 g for 30 min at 4°C in 50-ml tubes. Decant supernatant (containing MGF).
7. Calculate protein concentration (A_{280}) and store as a lyophilized powder at -20°C in 100-mg aliquots in plastic scintillation vials.
8. Reconstitute in 10 ml medium 199 for dialysis. Wash dialysis tubing (Mw cut off 6-8,000) extensively with ddH_2O. Transfer resuspended extract into dialysis tubing with Pasteur pipette and seal tubing. Dialyse overnight in 500 ml 200 mM Tyris-HCl buffer, pH 7.2, at 4°C. Replace buffer next morning with 500 ml fresh Tris-HCl and continue to complete 24-hr dialysis.
9. Filter sterilize with a 0.22-mm filter.
10. Using Tris-HCl as a blank, obtain A_{280} for a 1:10 dilution of the dialyzed extract. Bring concentration to 2 mg per ml with medium 100. Store at 4 °C.

Protocol: Establishing the primary culture

Day 1. Rinse skin specimen in 70% ethanol, then twice in PBSA. Transfer tissue to a sterile 100-ml dish, epidermal-side-down. With dissecting scissors, excise subcutaneous fat and deep dermis. Cut remaining tissue into 5 × 5-mm pieces with a scalpel, rolling blade over tissue. Transfer tissue fragments to cold 0.25% trypsin in a 15-ml centrifuge tube. Incubate tube at 4°C for 18-24 hr.

Day 2. Gently tap centrifuge tube to dislodge fragments settled in the bottom, then rapidly pour tube contents into a 60-mm dish. Using forceps, transfer tissue fragments individually to a dry 100-mm dish, epidermal-side-down. Gently roll each tissue fragment against the dish. The epidermal sheet should adhere and allow clean separation of the dermis with forceps. Discard dermal fragments. Transfer all epidermal sheets to a sterile 15-ml centrifuge tube containing 5 ml 0.02% EDTA, taking care to place each epidermal sheet in the EDTA solution, not on the plastic wall of the centrifuge tube. Gently vortex tube to disintegrate epidermal sheet in the EDTA solution, not on the plastic wall of the centrifuge tube. Gently vortex tube to disintegrate epidermal sheets into a single cell suspension. Centrifuge cells 5 min at 350 g, then aspirate supernatant. Resuspend pellet in serum-free melanocyte medium. Determine cell count using a hemocytometer and inoculate 10^6 cells (approximately 2-4 × 10^4 melanocytes) per 35-mm dish in 2 ml of hormone supplemented medium containing 2% serum to facilitate attachment. Reefed fresh serum-free medium twice weekly.

Days 3-30. At 24 hr, cultures will contain primary keratinocytes with scattered melanocytes. Keratinocyte proliferation should cease within several days and colonies should begin to detach during the second week. By the end of the third week, only melanocytes should remain. In most cases, cultures attain near confluence and are ready to passage within one month.

Subcultivation

1. Gently rinse culture dish twice with 0.02% EDTA. Add 1 ml 0.25% trypsin/0.1% EDTA and incubate at 37°C. Examine dish under phase microscopy every 5 min to detect cell detachment.
2. When most cells have detached, inactive trypsin with soybean trypsin inhibitor or 1 ml medium containing 10% serum. (Melanocytes maintained under serum-free conditions greatly benefit from a "serum kick" at the time of cultivation).
3. Pipette dish contents to ensure complete melanocyte detachment. Aspirate and centrifuge for 5 min at 350 g. Aspirate supernatant, resuspend cells in melanocyte medium containing 2% serum, and replate at 2-4 × 10^4 cells per 35-mm dish. Serum is required for good melanocyte attachment (> 75%) to plastic dishes, but can be omitted if dishes are coated with fibronectin or type I/III collagen. Refeed twice weekly with serum-free or serum-containing melanocyte medium, dependent on experimental protocol.

Confirmation of melanocytic identity

Melanocyte cultures may be contaminated initially with keratinocytes and at any time by dermal fibroblasts. Both forms of contamination are rare in cultures established and maintained by an experienced technician/investigator, but are common problems for the novice, Melanocytic identity of the cultured cells can be confirmed with moderate certainty by frequent examination of cultures under phase microscopy, presuming familarity with the respective cell morphologies. More definitive identification is provided by electron microscopic examination, dopa staining, or immunofluorescent staining with S 100 antibody.

Hemopoietic Cells

There have been three major milestones in this area. Bradley and Metcalf (1966), Pluznik and Sachs (1965), and McCulloch and co-workers developed techniques for cloning normal hemopoietic precursor cells in agar or Methocel in the presence of colony stimulating factor(s). The colonies matured during growth and could not be subloned, implying that the colony-forming unit (CFU) was a precursor cell which was not regenerated in culture. This cell, the CFU-C ("colony-forming unit-culture") is distinct from the CFU-S (spleen colony-forming unit), which is a pluripotent stem cell present in colonies forming in the spleens of sublethally irradiated mice after bone marrow reconstitution.

Hence, suspension colonies, which contain cells of only on lineage, survive only as primary cultures which lose repopulation efficiency and cannot be subcultured. Granulocytic colonies are the most common; but under the appropriate conditions, lymphoid and erythroid colonies can be produced.

Golde and Cline (1973) obtained survival in a liquid culture system of normal and neoplastic leukocytes at high cell densities but with abundant medium by placing the cells in a small diffusion chamber immersed in medium. Dexter has also demonstrated, in a liquid culture system, that lymphoid, granulocytic, and erythroid stem cells could be propagated from bone marrow if a bone marrow culture was first prepared and allowed to form a monolayer and to act as a feeder layer for a later, second bone marrow primary culture.

The third major development which occurred over several years between the earlier suspension cloning and Dexter's liquid culture system was the development of a number of functional cell lines from hemopoietic cells. Human lymphoblastic cell lines in both B and T cell lineage were developed by Moore et al. (1967) and subsequently Epsterin-Barr virus has been found to be implicated in the ability of these cell lines to become permanent. A number of myeloid cell lines have also been developed from murine leukemias and, like some of the human lymphoblastoid lines have been shown to make flobulin chains, and in some cases, complete α-and γ-globulins. Some of these lines can be grown in serum-free medium. T-cell lines require T-cell growth factors (e.g. Interleukin IL-2) and B-cell growth factors have also been described.

Originally human lymphoblastoid cell lines were derived by culturing peripheral lymphocytes from blood at very high cell densities (- 106/ml), usually in deep culture (> 10 mm). A monolayer culture appeared in the cell pellet at the bottom of the culture tube and eventually cells were shed into suspension and started to proliferate. This could be detected by the pH drop and the cells were then subcultured. The cell concentration was kept high initially, but eventually these cells adapted to regular culture conditions and could be passaged at 10^5 cells/ml or less. More recently the development of cell lines has become easier by the use of irradiated spleen cells, antigenic stimulation (for T-cell lines), and T- and B-cell growth factors.

Erythroid cell lines have also been cultured from the mouse. Rossi and Friend (1967) demonstrated that a mouse RNA virus (the "Friend virus") could cause splenomegaly and erythroblastisis in infected mice. Cell cultures taken from minced spleens of these animals could, in some cases, give rise to continuous cell lines of erythroleukemia cells. All of these cell lines are transformed by what is now recognized as a complex of defective and helper virus derived from Moloney sarcoma virus. Some cell lines can produce virus which is infective *in vivo* but not *in vitro*, and the cells can also be passaged as solid tumors or ascites tumors in DBA2 or BALBC mice.

Treatment of cultures of Friend cells with a number of agents, including DMSO, sodium butyrate, isobutyric acid, and hexamethyl-bis-acetamide, promotes erythroid differentiation. Untreated cells resemble undifferentiated proerythroblasts while treated cells show nuclear condensation, reduction in cell size, and an accumulation of hemoglobin to the extent that centrifuged cell pellets are red in colour. Evidence for differentiation can also be demonstrated by staining for hemoglobin with benzidine, isolating globin-specific messenger RNA, and fluorescent antibody detection of spectrim, a specific cell surface constituent of erythrocytes, on the surface of stimulated cells.

Anderson et al. (1979) have shown that the human leukemic cell line K562 can also be induced to differentiate with sodium butyrate and hemin, though not with DMSO.

Macrophages may be isolated from many tissue by collecting the cells that attach during enzymatic disaggregation. The yield is rather low, however, and a number of techniques have been developed to obtain larger number of macrophages. Mineral oil or thioglycollate broth may be injected into the peritoneum of a mouse, and 3 days later the peritoneal washing contain a high proportion of macrophages.

If necessary, macrophages may be purified by their ability to attach to the culture substrate in the presence of proteases, as above. They can only be subcultured with difficulty because of their insensitivity to trypsin. Methods have been developed using hydrophobic plastic, e.g., Petriperm dishes (Heraeus).

There are some reports of propagated lines of macrophages mostly from murine neoplasia. Normal mature macropahges do not proliferate although it may be possible to culture replicating precursor cells by the method of Dexter.

The following protocol for the long-term culture of bone marrow has been contributed by T.M. Dexter.

Principle

By culturing whole bone marrow the relationship between the stroma and cells is maintained and, in the presence of the appropriate hemopoietic cell and stromal cell interactions, proliferation of stem cells and specific progenitor cells can be maintained over several weeks. Progenitor cells from fresh marrow or long-term cultures may be assayed by clonogenic growth in soft agar or in mice.

Outline

Marrow is aspirated into growth medium and maintained as an adherent cell multilayer for at least 12 week and up to 30 week. Stem cells, maturing, and mature cells are released from the adherent layer into the growth medium. Granulocyte/macrophage progenitor cells can be assayed in soft gels.

Materials

(All reagents must be pretested to check their ability to support the growth of the cultures)

Fischer's medium (Gibco) supplemented with 50 U/ml penicillin and 50 μg/ml streptomycin and containing 16 mM (1.32 g/l) $NaHCO_3$.

Growth medium: as above, 100-ml aliquots supplemented with 10^{-6} M hydrocortisone sodium succinate and 20% horse serum (Flow Labs) (hydrocortisone sodium succinate made up as 10^{-3}M stock in Fischer's medium and stored at –20°C)

Five mice-strain differences exist

1-ml sterile syringes with 21-gauge needles

Gauze, swabs, scissors, forceps

25-cm² tissue culture flasks.

Protocol: Long-term bone marrow cultures

1. Kill donor mice by cervical dislocation. Wet the fur with 70% alcohol, remove both femurs. Collect ten femurs in a petri dish

on ice containing Fischer's medium. One femur contains 1.5-2.0 $\times 10^7$ nucleated cells.

2. In a laminar flow cabinet, clean off any remaining muscle tissue using gauze swabs. Hold the femur with forceps and cut off the *knee* end. The 21-gauge needle should fit snugly into the bone cavity. Cut off the other end of the femur as close to the end as possible. Insert the tip of the bone into a 100-ml bottle of growth medium and aspirate/depress the syringe plunge several times until all the bone marrow is flushed out of the femur. Repeat with the other nine bones.
3. Disperse to a fine suspension by pipetting the large marrow cores through a 10-ml pipette. There is not need to disaggregate small cell clumps. Dispense 10-ml aliquots of the cell suspension into 25-cm^2 tissue culture flasks, swirling the suspension often to ensure even distribution of the cells in the ten cultures. Gas the flasks with 5% CO_2 in air and tighten the caps. Incubate the cultures horizontally at 33°C.
4. Feed the cultures weekly. Agitate the flasks *gently* to suspend the loosely adherent cells. Remove 5 ml of growth medium including the suspension cells, taking care not to touch the layer of adherent cells with the pipette. Add 5 ml of fresh growth medium to each flask; to avoid damage do not dispense the medium directly onto the adherent layer. Gas the cultures and replace in the incubator.

Analysis

Cells harvested during feeding can be investigated by a range of methods including morphology, CFC-assays, and *in vivo* CFU-S assay for stem cells.

Variations

Human long-term cultures have been grown.

GM-CFC

The following protocol for GM-CFC Assays has also been contributed by T.M. Dexter.

Principle

Hemopoietic progenitor cells may be cloned in suspension in semisolid media, in the presence of the appropriate growth factor(s). Pure or mixed colonies will be obtained depending on the potency of the stem cells isolated.

Outline

Fresh bone marrow or supernatant cells from long-term cultures are diluted in a growth factor enriched planting medium, mixed with melted agar, and plated out. Colonies form in suspension in the gelled agar.

Materials

Fischer's medium, horse serum, mice, syringes, 21-gauge needles, scissors, swabs, and forceps

Granulocyte/macrophage colony stimulating factor (GM-CSF) or interleukin 3 (IL-3). GM-CSF is prepared by conditioning medium for 2 d with mouse lung tissue and IL-3 by conditioning medium with WEHI-3B cells for 3-6 d. Both these growth factors have been molecularly cloned and recombinant material will soon be commercially available.

3.3% Noble agar (Difco) in double distilled water sterilized by boiling

White cell diluting fluid (WCFD): 3% glacial acetic acid (nonsterile) coloured with gentian violet

Sterile bottles in which to mix plating medium: capacity > 6 ml

35-mm tissue culture grade plastic petri dishes.

Protocol: GM-CFC assay

1. Flush the marrow cells into Fischer's medium as described in steps 1 and 2 above. Count the nucleated cells in a hemocytometer using WCFD and adjust to the appropriate concentration. Fresh bone marrow cells should be plated at about 5 × 10^4 ml, 2.5 ml. Warm the plating mixture to above 20°C.
2. The plating mixture is made as follows: (× 10 final desired concentration), 0.5 ml; horse serum (final concentration 20% v/v), 1.0 ml; WEHI-CM 10% v/v or lung-CM (final concentration 10% v/v), 0.5 ml; Fischer's medium to make final volume to 5.0 ml, 2.5 ml. Warm the plating mixture to above 20°C.
3. Melt the agar in a boiling waterbath and add 0.5 ml per plating mixture.
4. Rapidly mix the plating mixture thoroughly by pipetting to ensure the agar is evenly distributed. Dispense 1-ml aliquots into triplicate 35-mm petri dishes. Swirl the plates gently so the agar mixture covers the base of the plate.

5. Place the agar cultures in the refrigerator for about 5 min to set the agar.
6. Place the cultures at 37°C in a humidified atmosphere of 5% CO_2 in air.
7. After 7 d growth the colonies should be ready for scoring. Colonies can be scored using a steremicroscope (magnification about 25). A colony is classed as a group of more than 50 cells. It may have a very compact form, a diffuse form, or it may have a compact center with a diffuse halo. These colony types are likely to be granulocytic, monocytic, and mixed granulocyte/ monocyte respectively. However, in order to classify colony types correctly, the colonies must be picked out, disaggregated, cytocentrifuged onto slides, and stained.

Variations

CFC assay exist for mouse multipotent-CFC, erythroid-CFC, megakaryocyte-CFC, B-cell-CFC, eosinophil-CFC, and "fibroblast"-CFC (a component of the hemopoietic environment). A similar range of human CFC can also be grown.

Gonads

Culture of germ cells has on the whole been disappointing. Ovarian granulosa cells can be maintained and are apparently functional in primary culture, but specific functions are lost on subculture. A cell line started from *Chinese hamster ovary* (CHO-K1) has been in culture for many years but its identity is still not confirmed. Although epithelioid at some stages of growth, it undergoes a fibroblastic-like modification when cultured in dibutyryl cyclic AMP.

Cellular fraction from testis have been separated by velocity sedimentation at unit gravity, but prolonged culture of these has not been reported. The TM4 is an epithelial line from mouse testis although its differentiated features have not been reported, and Sertoli cells have also been cultured from testis.

Minimal Deviation Tumors

Several cell lines have been derived from the Reuber and Morris hepatomas of the rat adrenal cortex and pituitary (as described above) and provide a valuable, if rare, source of continuous cell lines with differentiated properties.

Teratomas

When cells from an embryo are implanted into the adult, e.g., under the kidney capsule, these can give rise to tumors known as *teratomas*. Teratomas also rise spontaneously when groups of embryonic cells or single cells are carried over into the adult, often at an inappropriate site.

Artificially derived teratomas have been used extensively to study differentiation, as they may develop into a variety of different cell types (muscle, bone, nerve, etc). Growth of teratoma cells on feeder layers of, for example, SCI mouse fibroblasts, will proliferate but not differentiate, whereas when grown on gelatin without feeder layer, or in nonadherent plastic dishes, nodules form which eventually differentiate.

17

CELL POPULATION

An important growth parameter used for cultured cells is the population assay. It is a good index from which to establish the best culture conditions for a cell line. To determine a growth curve, cells are plated at low densities, and are counted at fixed time intervals, without feeding or diluting the culture. The number of cells in the culture is plotted against the elapsed time, and this data generates a curve which can be divided into four stages.

(a) *Lag phase*. The first stage of a growth curve is called the *lag phase*. In this phase of growth, the cells grow slowly or not at all. At this stage cells adapt to new conditions such as type of medium, serum concentration, type of culture vessel and initial cell density. These factors determine the length of the lag phase.

(b) *Log phase*. The next stage of a growth is called *log phase*. Here the cells will grow exponentially and will divide as fast as possible. In this stage, the medium is rich in nutrients, and there is still sufficient space in the culture vessel available for the cells to grow without inhibiting or competing with each other. It is during the log phase that the doubling time of a culture should be determined. Doubling time is the time it takes for a culture in log phase to increase by a factor of two.

(c) *Plateau phase*. The next stage in growth curve is called the *plateau phase*. Here the cells begin to exhaust the available nutrients and growing space, and their rate of division slows down. At this stage, the maximum number of cells which can be grown per unit volume of medium can be determined. This saturation density represents the density at which the cells can no longer grow exponentially and, it too, will vary depending on the conditions used.

(d) *Death phase*. The last stage of the curve is called the *death phase*. With the depletion of nutrients and accumulation of waste products, the cells will begin to die.

Safety Guidelines

Wear gloves, lab coat, and safety glasses. Standard laboratory safety procedures should be followed.

Materials

1. A bottle of complete Fischer's medium (Fischer's medium supplemented with 10% v/v with horse serum, plus 1 mM sodium pyruvate).
2. Culture of mouse L-5178Y cells or any line that grows In suspension (note that other cell lines may require different media)
3. Concentrated trypan blue solution - 0.4%.
4. CO_2 incubator or a tank of 5% CO_2.

Method

1. Obtain a culture of L-5178Y cells growing in 25 cm^2 flask in Fisher's medium.
2. Tighten the cap of the flask and shake it well from side to side.
3. Using a 10 ml pipette, aspirate cell suspension up into and out of the pipette 10 times to break up the clumps of cells.
4. Transfer 0.1 ml of the cell suspension to a 1.5 ml microcentrifuge tube with attached cap. Close the culture flask with its cap.
5. Add 0.1 ml trypan blue solution and count the cells using a hemocytometer. Trypan blue is used to check viability.
6. Record the cell count In the original cell suspension. Make proper dilution so that it now contains 2×10^4 cells/ml in about 20 ml of the medium. This is the culture that will be used for generating growth curve.

Using the formula $V_1C_1 = V_2C_2$ calculate what volume of the original cell suspension (V_1) when diluted to a final volume of 20 ml (V_2) yields a final density of 2×10^4 cells/ml (C_2). If the original density (C_1) of your cell suspension is 1×10^6 cells/ml, you will set up the formula as follows:

$$V_1\ (1 \times 10^6) = (20\ \text{ml})\ (2 \times 10^4)$$

$$\frac{V_1 = (20\ \text{ml} \times 2 \times 10^4)\ \text{cells}}{1 \times 10^6\ \text{cells}}$$

$$= \frac{40 \times 10^4}{1 \times 10^6} = \frac{4 \times 10^5}{1 \times 10^6} = \frac{4}{10} = 0.4 \text{ ml}$$

7. Transfer this volume into a graduated sterile cylinder and bring the volume to 20 ml using Fisher's medium. Transfer this to 25 cm^2 culture flask and cap it. This culture will be used for generating a growth curve. Count this culture at least twice a day.
8. Place the flasks upright in the CO_2 incubator at 37°C. Make sure that the cap is loosened before returning the cells to the CO_2 incubator. Alternatively, gas the culture with 5% CO_2 and keep the cap tightened. Remember that the concentration of cells at 0 time is 2×10^4 cells/ml.
9. At least two measurements should be taken on the first day of culture. For example, at about three and nine hours after time zero are ideal.
10. Note the time of day when each set of daily measurements are taken and calculate the age of the culture, in hours from time zero, for each set of density determinations. Keep a written record of all cell counts. After each set of determinations, return the flasks to the CO_2 incubator, with caps loosened or to the regular incubator, with caps tightened following a 10-15 second bubbling of 5% CO_2.
11. Continue this procedure until the medium has turned yellow and the density of viable cells has declined considerably.

Results

Once the data has been collected, construct a graph using semilog paper by placing the time in hours on the x-axis and the cell count on the y-axis.

18

Three-dimensional Culture Systems

When tissue culture was first developed it was based on the explanation of whole fragments of tissue with a view to studying them as tissue in isolation. However, it was observed that cells often grew out from these tissue in a sheet to form a monolayer on the supporting glass substrate. This divergence in growth properties–growth as migratory and potentially proliferative monolayer versus residence within the original explant–set the pattern for future divergence in approach to the culture of animal cells. One school of thought believed in the retention of histological structure and possibility of organotypic function while the other looked towards the biology of individual cells. The former required retention of histotypic structure, cell interaction, and histological characteristics and enabled the study of developmental problems such as embryonic induction, *in vitro* modeling of malignant invasion, and hormonal control of morphogenesis and differentiation. The latter gave rise to propagated cells lines which laid down the basis for most of modern cellular and molecular biology and its insight into the regulation of gene expression.

Now, although the potential for propagated cell lines is far from exhausted, many people are reverting to the notion that nutritional completeness and hormonal supplementation are inadequate in themselves to recreate full structural and functional competence in a given cell population. The vital missing factor is cell interaction and the signaling capacity that it entails. A hormone may activate a specific pathway in cell A, and it or a different hormone activate a different commitment in cell B. Both, alone, may lead to individual modifications in phenotypic expression, but if allowed to interact, a cascade of interactions may occur dependent on the cells being in association. Alveolar cells of the lung will only synthesize release surfactant in response to hormonal stimulation of adjacent fibroblasts prostate epithelium response to stromal signals is in turn activated by androgen binding to the stroma. Epithelium differentiates in response to matrix constituents often determined jointly by the epithelium on one side and connective tissue stroma on the other, as may be the case with the interaction between epidermis and dermis *in vitro*. Hence the whole integrated tissue may easily, and understandably, respond differently to simple ubiquitous signals, not because of the specificity of the signal or the receptor capacity but because of the response encoded in the juxtaposition of one cell type with a specific correspondent. As in human society, the response of one individual to an exogenous stimulus will be dictated as much by the spatial and temporal relationship with other individuals as by the endogenous make-up. Likewise a primitive neural crest cell may become a neurone, an endocrine cell, or a teratoma, dependent on its ultimate location, its interaction with adjacent cells and its response, mediated by neighbouring cells, to hormonal stimuli.

In essence this preamble establishes that while some cell functions, such as cell proliferation, glycolysis, respiration, and gene transcription, proceed in isolation, their regulation, as related to a functioning multicellular organism, must depend ultimately on the interaction between cells of the appropriate lineage, the appropriate stage in that lineage, and in the appropriate intracellular regulatory environment. This suggests that if you want to study cell biology cell lines are fine, but if you want to learn something of the integrated function, or dysfunction, of whole organisms a histotypic or organotypic model will be required.

There are two major ways to approach this goal. One is to accept the cellular distribution of the tissue, explant it, and maintain it as an organ culture. The second is to purify and propagate individual cell lineages, recombine them, and study their interactions. This has given rise to three major types of technique:

Organ culture, where whole organs, or representative parts, are maintained as small fragments in culture and retain their intrinsic distribution, numerical and spatial, or participating cells; *histotypic culture*, where propagated cell lines are grown to high density in a 3D matrix alone; or *organotypic culture*, in which cells of different lineages are recombined in experimentally determined ratios and spatial relationships to recreate a component of the organ under study.

Organ culture seeks to retain the original structural relationship of different and similar cells and hence their interactive function, in order to study the effect of exogenous stimuli on further development. This may even be achieved by separating the constituents and recombining them as in the now classical experiments of Grobstein and Auerbach and others in organogenesis. Organotypic culture represents the synthetic approach whereby a three-dimensional, high-density culture is regenerated from isolated (and preferably purified and characterized) lineages of cells that are then recombined, their interaction studies, and, in particular their response to exogenous stimuli characterized.

"Exogenous stimuli" may be regulatory hormones, nutritional conditions, or environmental toxins. In each case the response, and the justification of this approach, will be different from the responses of a pure cell type in isolation.

Several types of systems have been described to study isolated, whole, undisaggregated tissue or recombinations of tissues, or purified cell lineages. As these may provide models for quite distinct types of investigation each will be described separately.

ORGAN CULTURE

Gas and Nutrient Exchange

When cells are cultured as a solid mass of tissue, gaseous diffusion and the exchange of nutrients and metabolites becomes limiting. The dimensions of individual cells cultured in suspension or as a monolayer are such that diffusion is rapid but aggregates of cell beyond about 250 μm (5,000 cells) start to become limited by diffusion and at or above 1.0-mm diameter ($\sim 25 \times 10^5$ cells) central necrosis is often apparent. To alleviate this problem organ cultures are usually placed at the interface between the liquid and gaseous phase to facilitate gas exchange while retaining access to nutrients. Most systems achieve this by positioning the explant on a raft or gel exposed to the air but explants anchored to a solid substrate can also be aerated by rocking the culture, exposing it alternately to liquid medium or gas phase or by using a roller bottle or tube to the same end.

Anchorage to a solid substrate can lead to the development of an outgrowth from the explant and resultant alterations in geometry although this can be minimized using a nonwettable surface. One of the advantages of culture at the gas-liquid interface is that the explant retains a spherical geometry if the liquid is maintained at the correct level: too deep and gas exchange is impaired; too shallow and surface tension will tend to flatten the explant and promote outgrowth.

Increased permeation of oxygen can also be achieved by using increased O_2 concentrations up to pure oxygen or by using hyperbaric oxygen. Certain tissues, e.g. thyroid and prostate trachea and skin particularly from newborn or adult may benefit from elevated O_2 tension but often this is at the risk of O_2 induced toxicity. This may have to be determined for each tissue type under study.

Increasing the O_2 tension will not facilitate CO_2 release or nutrient metabolite exchange, so the benefit of increased oxygen may be overridden by other limiting factors.

Structural Integrity

Structural integrity, above other considerations, was and is the main reason for adopting organ culture as *in vitro* technique in preference to cell culture. While cell culture utilizes cells dissociated by mechanical, enzymic techniques or spontaneous migration, organ culture deliberately maintains the cellular associations found in the tissue. Initially this was selected to facilitate histological characterization but ultimately it was discovered that certain elements of phenotypic expression were only found if cells were maintained in close association.

It is now recognized that associated cells do exchange signals via junctional communications ("gap" junctions) and via paracrine hormones and surface information exchange. This is most striking during organogenesis, but is probably also required for the maintenance of fully mature tissues.

Hence maintenance of the structural integrity of the original tissue may preserve the correct homologous and heterologous cellular interactions present in the original tissue, and maintain the correct chemical configuration of the extracellular matrix.

A major deficiency in tissue architecture in organ culture is the absence of a vascular system, limiting the size and potentially also the polarity of the cells within the organ culture.

Growth and Differentiation

It has been stated previously that there appears to be relationship between growth and differentiation such that differentiated cells no longer proliferate. It is also possible that cessation of growth may in itself contribute to the induction of differentiation if only by providing a permissive phenotypic state receptive to exogenous inducers of differentiation.

Because of the rules of density limitation of growth, and the physical restrictions imposed by their geometry, most organ cultures do not grow, or if they do, proliferation is limited to the outer cell layers. Hence the status of the culture is permissive to differentiation and given the appropriate cellular interactions and soluble inducers should provide an ideal environment for differentiation to occur.

Limitations of Organ Culture

In view of the advantages described above it is perhaps surprising that organ culture is not more popular. The reasons relate largely to the development of biochemical and molecular criteria for *in vitro* behaviour, particularly the monitoring of differentiation. These criteria have been adopted because they are more readily quantified and generally more objective than histological criteria although they lack the resolution of histological techniques where local response in a minority of cells can be detected.

Biochemical monitoring requires reproducibility between samples which is less easily achieved in organ culture than in propagated cell lines. This is due to the sampling variation in preparing an organ culture, to monitor differences in handling and geometry, and to variations in cell type heterogeneity between cultures.

Organ cultures are also more difficult to prepare than replicate cultures from a passaged cell line and do not have the advantage of a characterized reference stock to which they may be related. Preparation is labour intensive and as a result the yield of usable tissue is often too low to be a value in biochemical or molecular assays. Furthermore as the reacting cell population may be a minor component it is difficult to analyze the biochemical nature of the response and attribute it to the correct cell type other than by autoradiographic, histochemical, or immunocytochemical techniques that tend to be more qualitative than quantitative.

Organ cultures cannot be propagated and hence each experiment requires recourse to the original donor tissue.

Organ culture is essentially a technique for studying the behaviour of integrated tissue rather than isolated cells. It is precisely in this area that the future understanding of the control of gene expression (and ultimately of cell behaviour) in multicellular organisms may lie, but the limitations imposed by the organ culture system are such that recombinant systems between purified cell types may contribute more at this particular stage. However, there is no doubt that organ culture has contributed a great deal to our understanding of development biology and tissue interactions and that it will continue to do so in the absence of adequate synthetic systems.

Types of Organ Culture

As the technique has been dictated largely by the requirement to place the tissue at a location allowing optimal gas and nutrient exchange, most techniques place the tissue at the gas-liquid interface.

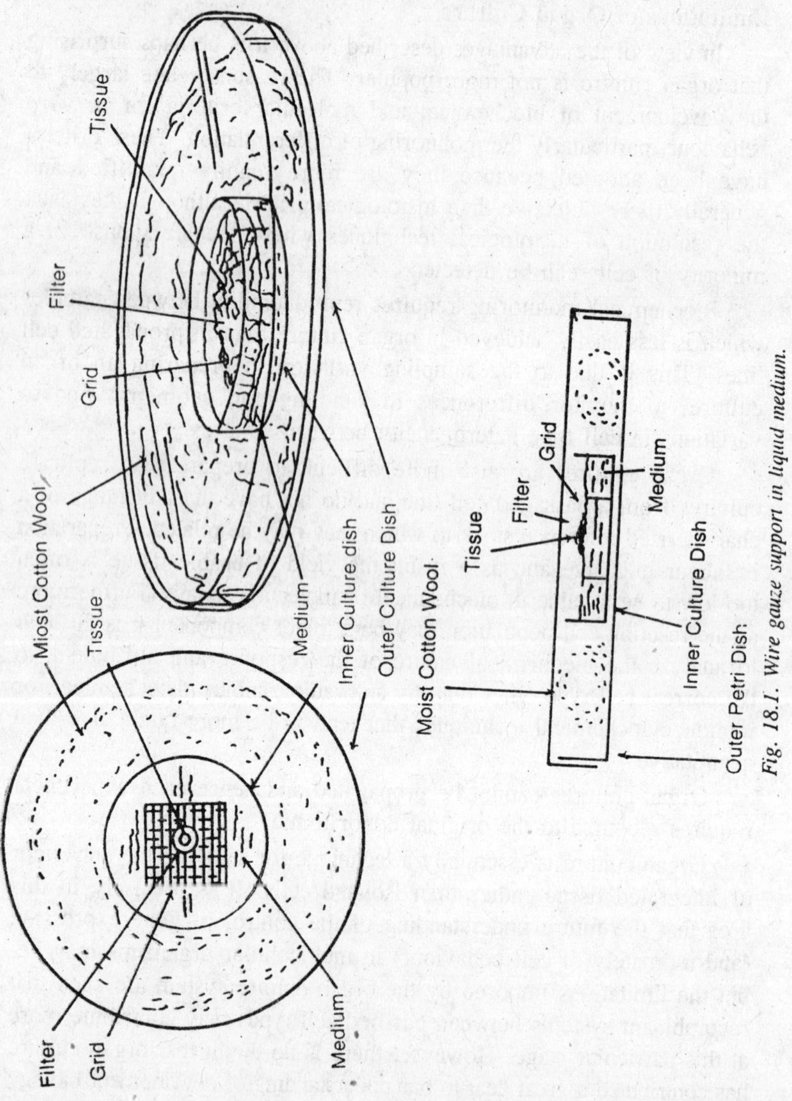

Fig. 18.1. Wire gauze support in liquid medium.

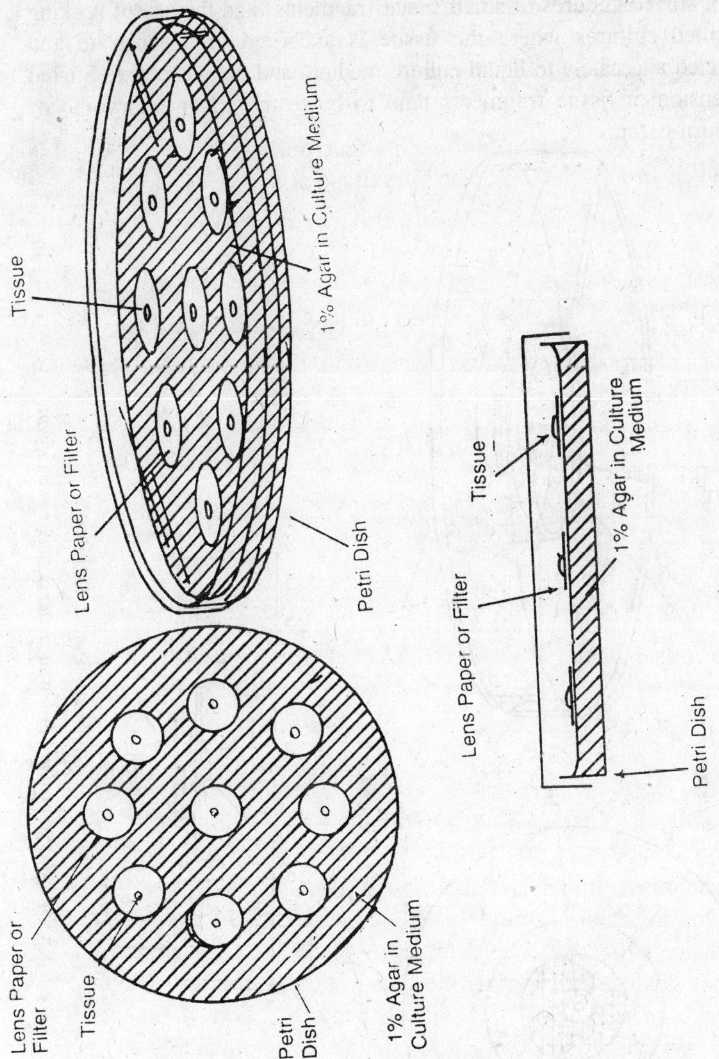

Fig. 18.2. On filter or lens paper rafts on agar medium.

This has been achieved by placing fragments of tissue on semisolid gel substrates of agar or clotted plasma or on a raft of microporou; filter, lens paper, or rayon supported on a stainless steel grid or adherent to a strip of Perspex or Plexiglas.

Two techniques have successfully departed from this method, the use of stirred cultures of small tissue fragments, and the use of rocking or rolled cultures where the tissue is anchored to a substrate and subjected alternately to liquid culture medium and the gas phase. Stirred suspension of tissue fragments tend to be restricted to embryonic or newborn tissue.

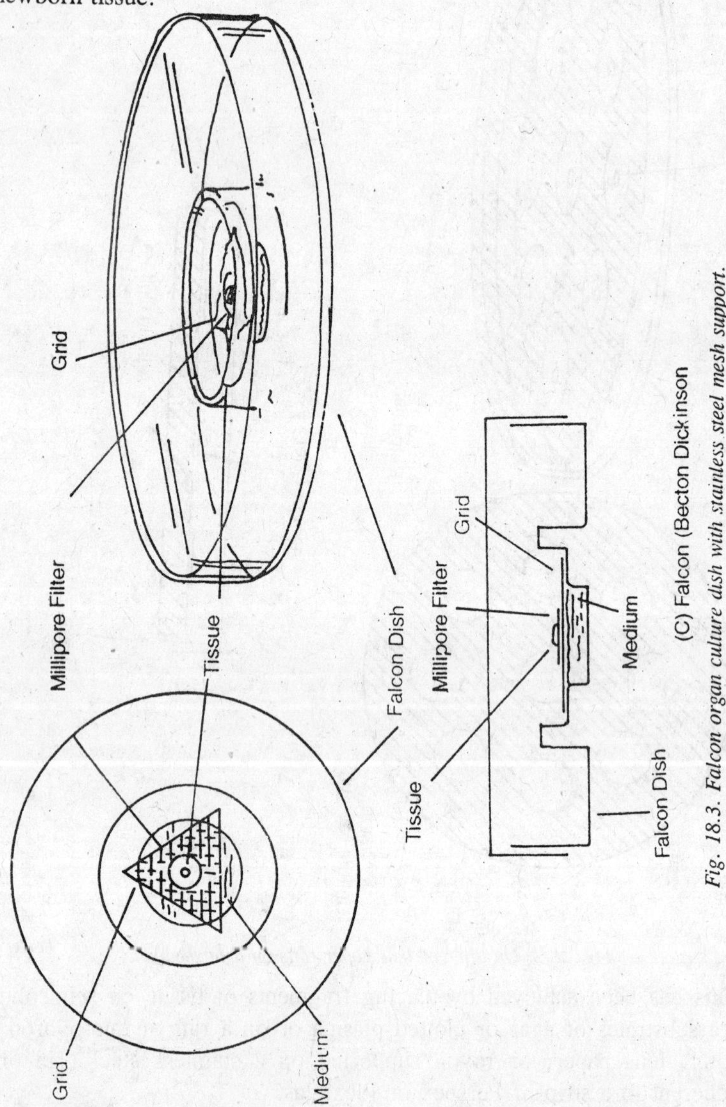

Fig. 18.3. Falcon organ culture dish with stainless steel mesh support.

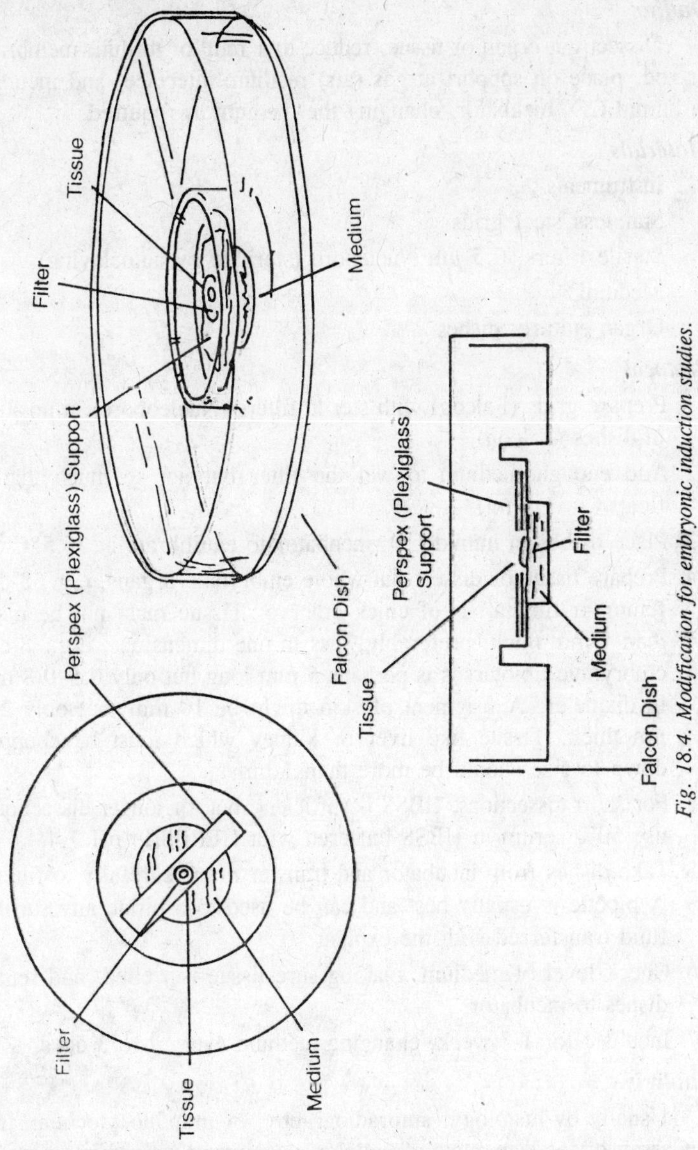

Fig. 18.4. Modification for embryonic induction studies.

Several techniques have been described for gas liquid interface culture, but one which has become popular because of the provision of suitable disposable plasticware is the grid technique derived from that of Trowell (954, 1559).

Outline

Dissect out organ or tissue, reduce to 1 mm^3 or to thin, membrane or rod, place on support at gas (air) medium interface, and incubate in humid CO_2 incubator, changing the medium as required.

Materials

Instruments

Stainless steel grids

Sterile filters, 0.5 μm Nuclepore (sterilize by autoclaving)

Medium

Organ cultures dishes

Protocol

1. Prepare grids (Falcon) with sterile filters (Nucleopore) in position in dishes (Falcon).
2. Add enough medium to wet the filter but not so much that if floats ($\sim$ 1.1 ml).
3. Place dishes in humid CO_2 incubator to equilibrate at 36.5°C.
4. Prepare tissue or dissect out whole embryonic organs, e.g., 8 day femur or tibiotarsus of chick embryo. Tissue must not be more than 1 mm thick, preferably, less in one dimensions, e.g., 8-day embryonic tibiotarsus is perhaps 5 mm long but only 0.5–0.8 mm in diameter. A fragment of skin might be 10 mm^2 but only 200 μm thick. Tissue like liver or kidney which must be chopped down to size should be more than 1 mm^3.

   For short dissections, HBSS is sufficient, but for longer dissections, use 50% serum in HBSS buffered with HEPES to pH 7.4.
5. Take dishes from incubator and transfer tissue carefully to filters. A pipette is usually best and can be used to aspirate any surplus fluid transferred with the explant.
6. Check level of medium, making sure tissue is wetted, and return dishes to incubator.
7. Incubate for 1-3 week, changing medium every 2 to 3 days.

Analysis

Usually by histology, autoradiography, or immunocytochemistry, but assay of total amounts of cellular constituents or enzyme activity is possible, although variation between replicates will be high.

Variations

Most variations are in:

1. Type of medium; 199 or CMRL 1066 may be used with or without serum, and BJG for cartilage or bone.
2. Type of support. The Grobstein technique has a number of advantages. Different types of tissue may be combined on the opposite sides of the filter to study their interaction. Furthermore, the well formed on the top side of the filter assembly generates a meniscus of medium with a large surface area available for gas exchange. It is also possible to alter the configuration of the tissue by raising or lowering the level of medium in the dish, and thereby in the well; deeper medium gives a spherical explant and shallower medium flattens the explant. This system does have the disadvantage that the filter assemblies are not commercially available.
3. O_2 tension; embryonic cultures are usually best kept in air, but late-stage embryos, newborn, and adult tissue are better kept in 95% O_2:

Organ cultures are useful in the demonstration of processes such as embryonic induction where the maintenance of the integrity of whole tissue is important. However, they are slow to prepare and present problems of reproducibility between samples. Growth is limited by diffusion (although growth is perhaps not necessary and may even be undesirable) and mitosis is nonrandomly distributed throughout the explant. Mitosis occurs round the periphery only, while the centers of the explants frequently become necrotic. It has been argued that this type of geometry makes organ cultures good models of tumor growth, where peripheral cell division is often accompanied by central necrosis.

HISTOTYPIC CULTURE

Various attempts have been made to regenerate tissue like architecture from dispersed monolayer cultures. Kruse and Miedema (1965) demonstrated that perfused monolayers could grow to more than ten cells deep and organoid structures can develop in multilayered cultures if kept supplied with medium. Green (1978) has shown that human epidermal keratinocytes will form dematoglyphs (friction ridges) if kept for several weeks without transfer, and Folkman and Haudenschild (1980) were able to demonstrate formation of capillary tubules in cultures of vascular endothelial cell cultured in the presence of endothelial growth factor and medium conditioned by tumor cells.

Sponge Techniques

Leighton first demonstrated that cells would penetrate cellulose sponge. Both normal and malignant cells can do this and it does not seem to reflect malignant behaviour. Collagen coating of the sponge may facilitate occupation and Gelfoam (a gelatin sponge matrix used in reconstructive surgery) may be used in place of cellulose. These systems require histological analysis and are limited in dimensions, like organ cultures, by gaseous and nutrient diffusion.

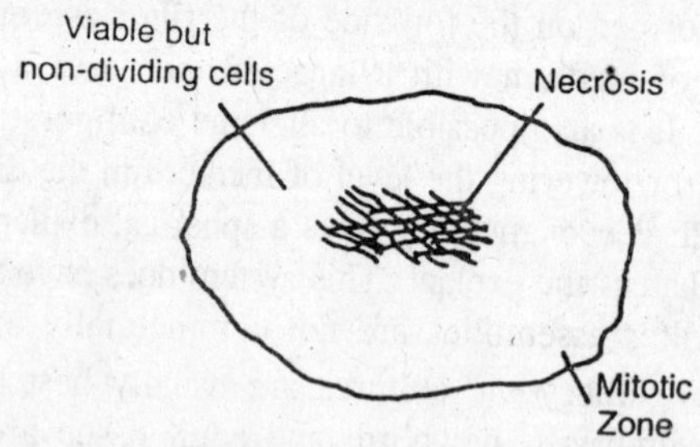

Fig. 18.5. Diagrammatic representation of the expected distribution of mitoses (stippled area) and necrosis (shaded central area) in an organ culture explant.

Collagen gel (native collagen as distinct from denatured collagen coating) provides a matrix for the morphogenesis of primitive epithelial structures. Yang et al. (1979, 1980, 1981) have shown that breast epithelium forms rudimentary tubular and glandular structures grown in collagen. Analysis is again by histology.

Capillary Bed Perfusion

Since medium supply and gas exchange become limiting at high cell densities, Knazek et. al. (1972); Gullino and Knazek, 1970) developed a perfusion chamber from a bed of plastic capillary fibers. This is now available commercially from Amicon. The fibers are gas- and nutrient-permeable and support cell growth on their outer surfaces. Medium, saturated with 5% CO_2 in air, is pumped through the centers of the capillaries, and cells are added to the outer chamber surrounding the bundle. The cells attach and grow on the outside of the capillary fibers fed by diffusion from the perfusate and can reach tissue-like cell densities. There is an option between two types of plastic and different ultrafiltration properties giving molecular weight cut-off points at 10,000, 50,000 or 100,000 daltons, regulating the diffusion of macromolecules from the medium to the cells.

It is claimed that cells in this type of high-density culture behave as they would *in vivo*. Choriocarcinoma cells release more human chorionic gonadotrophin and colonic carcinoma cells produce elevated levels of CEA. There are considerable technical difficulties in setting up the chambers, however, and they are costly.

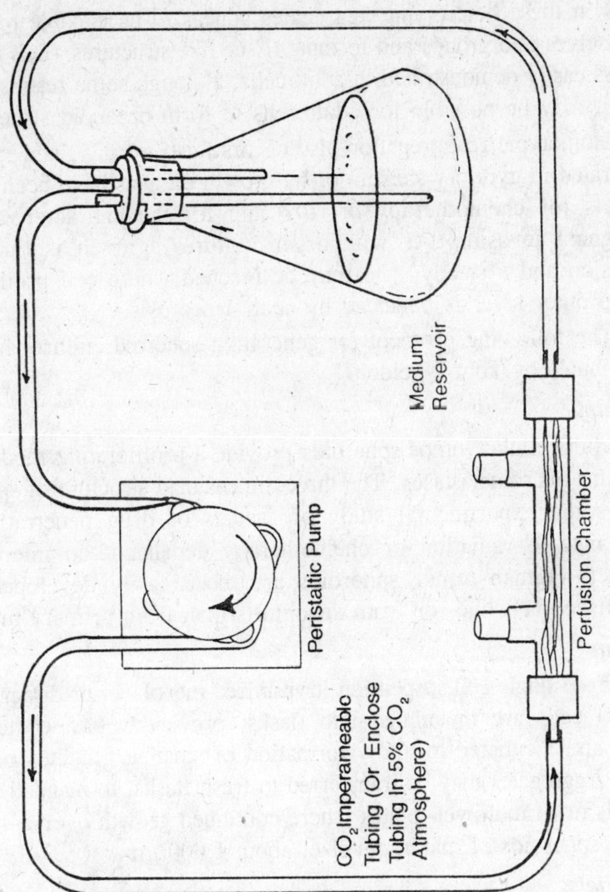

Fig. 18.6. **Vitafiber** *(Amicon) apparatus for perfused culture on capillary bundles of permeable plastic.*

Sampling cells from these chambers and determination of the cell concentration are also difficult. However, they appear to present in ideal system for studying the synthesis and release of biologically generated compounds and are now being exploited on a semi-industrial scale (Endotronics).

Reaggregation and Spheroids

When dissociated cells are cultured in a gyratory shaker, they may reassociate into clusters. Dispersed cells from embryonic tissues will sort during reaggregation in a highly specific fashion e.g., Muller cells of the chick embryo retina reaggregated with neuronal cells from the retina were inducible for glutamine synthetase; but those reaggregated with neurons from other parts of the brain were not.

Cells in these heterotypic aggregates appear to be capable of sorting themselves into groups and forming tissue-like structures. This property is less easily demonstrated in adult cells, although some results suggest that it may be possible for adult cells to form organoid structures.

Homotypic reaggregation also occurs fairly readily, and spheroids generated in gyratory shakers or by growth on agar have been used as models for chemotherapy *in vitro* and for the characterization of malignant invasion. As with organ cultures, growth is limited by diffusion and a steady state may be reached where cell proliferation in the outer layer is balanced by central necrosis.

The following protocol for generating spheroid cultures has been contributed by Tom Wheldon.

Principle

Multicellular tumor spheroids provide a proliferating model for a vascular micrometastases. The three-dimensional structure of spheroids allows the experimental study of aspects of drug penetration and resistance to radiation or chemotherapy dependent on intercellular contact. Human tumor spheroids are more easily developed from established cell lines or from xenografts than from primary tumours.

Outline

From single cell suspension (trypsinized monolayer or disaggregated tumor) cells are inoculated into flasks, previously base-coated with agar, and incubated to allow formation of small aggregates over 3-5 days. Aggregates may be transferred to fresh flasks, to magnetic stirrer vessels or to multiwell plates where continued growth over 2-4 weeks yields spheroids of maximum size, about 1,000 μm.

Materials

Noble agar (Difco)

Growth medium

Distilled water (sterile)

Trypsin 0.25% in PBS

25-cm^2 flalsks or 24 well plates.

9 cm petri dishes

Note : where agar coating is used flasks, plates, and dishes should be sterile but not necessarily tissue culture grade.

Protocol

1. *Agar coating.* In 25 cm^2 flasks: add 1g Noble agar (Difco) to 20 ml distilled water in a 150-ml Erlenmeyer flask and boil gently

until the agar has completely dissolved (about 10 min). Add contents immediately to 60 ml of growth medium, previously heated to 37°C, and put 5 ml aliquots in about 5 min giving a 1.25% agar coated flask. In multiwell plates: and 0.5 g agar to 10 ml distilled water, heat as above and then add 40 ml of distilled water. Place 0.5 ml of the resultant solution in each well of a 24-well plate (Corning 25820) to give a base coat of 1% agar. Accuracy and careful placement is important to ensure easy well-to-well focus of the microscope in subsequent viewing of spheroids.

2. *Spheroid initiation.* Trypsinize confluent monolayer (established lines) or disaggregate (solid tumors) to give single cell suspension. Neutralize trypsin with medium containing serum (if necessary). Count cells (Coulter counter or hemocytometer). Place 5×10^5 cells in each agar-coated 25 cm^2 flask in 5 ml growth medium and incubate. If the cells are capable of spheroid formation, small aggregated clumps (about 100-300 μM diameter) will form spontaneously in 3-5 days.
3. For subsequent growth, spheroids should be transferred to new vessels. If growth is to be continued in 25 cm^2 flasks, the contents of the original flask should be transferred to conical centrifuge tubes or universal containers, the spheroids allowed to settle, and single cells removed with supernatant. Spheroids may then be resuspended in fresh medium and transferred to new agar-coated flasks, where growth will proceed by divisions of cells in the outer layer.
4. For growth in wells, decant the contents of 25cm^2 flask into a 9 cm petri dish and, under laminar flow conditions, select individual spheroids of chosen dimensions under low power magnification ($\times$ 40). Using a Pasteur pipette and a Pi-Pump transfer selected spheroids of similar diameter individually to agar-coated wells or a 24 well plate, each containing 0.5 ml medium. Place plates in CO_2 incubator. Replace medium once or twice weekly (exchanging 0.5 ml each time) or add 0.5 ml medium (without removal) once or twice weekly giving 2-4 week for a 2ml well).
5. Spheroid growth in wells or flasks may be quantified by regular measurement (e.g., two to three times weekly) or diameter using a microscope graticule, or better, by measurement of cross-sectional area using an image analysis scanner. The most accurate growth curves are obtained when spheroids are grown in wells and individually monitored.

Variations

1. Some spheroids can be grown only when subjected to continuous agitation. Transfer single cells in suspension, or preformed aggregates, to siliconized glass culture vessels and agitate on a magnetic stirrer (Techne).
2. As a normal stromal cells are usually excluded from spheroids, spheroids formed from a disaggregated tumor cell suspension will usually contain only tumor cells. Monolayer cultures of tumor cells may then be established by placing spheroids in uncoated tissue culture petri dishes, to which they will attach. Monolayers will grow out from each attached spheroid and may be harvested by trypsinization.

Applications

Spheroids have wide applications in assessment of cytotoxic treatment. End-points include treatment induced growth delay, proportion of spheroids sterilized ("cured") by treatment, and colony formation in monolayer following disaggregation of treated spheroids. Some (not all) spheroid types may be "stripped" by exposure to proteolytic enzymes; exposure for varying periods followed by sequential removal of cells liberated allows assessment of cell survival or drug penetration in progressively deeper layers of the spheroid.

Filter Wells

Several attempts have been made to generate dense population of cells supported on a filter, either perfused from below by aerated medium or located at the gas-liquid interface or near to it. Mauchamp demonstrated the development of polarity and functional integrity in thyroid epithelium explanted on a collagen coated filter in a specially constructed mount. Others have used the system to study invasion by granulocytes or malignant cells.

One of the major advantages of the system is that it allows the recombination of cells at very high tissue like densities, with ready access to medium and gas exchange, but in a multireplicate form. McCall et. al., constructed chambers from disposable 2-ml syringes which utilized multiwell plates for support, and Millipore has now produced the Millicell-HA chambers in two sizes, 12 mm and 30 mm diameter suitable for 24 well plates, 6 well plates, or larger dishes. These filters are available in HA grade Millipore filters of different porosities but as yet are not transparent. The author has used transparent polycarbonate filters (Nucleopore) with defined and regular pores (1 μm or 8 μm are the most useful) in polypropylene mounts. These are useful for invasion studies and cell interacting studies but require further treatment to promote cell adhesion. Similar filters are produced by Coster, already mounted in polystyrene (Transwells).

Outline

Cells are seeded into filter chambers and cultured in excess medium in deep petri dishes or sample post (Sterilin, Macom).

Materials

Approximately 2 × 10^6 per filter
Growth medium, 20 ml per filter
Filter holders (Hendley Engineering)
Filter (polcarbonate 1-μm or 8-μm porosity, Nucleopore treated by Flow Laboratories–filter wells are assembled and sterilized by autoclaving, ethylene oxide, or γ-irradiation; autoclaving tends to distort the filter; must be left 2-3 weeks after ethylene oxide treatment before use; irradiation preferable (5 × 10^4 Gy, 5 Mrads)
Containers for filter assembly—Corning 9-cm petri dishes or deep sample posts (Sterilin or Macom).

Protocol

1. Place filter wells in dish or container.
2. Add medium, tilting dish to allow medium to occupy space below filter and displace air with minimum entrapment. Add medium until level with filter (15 ml in 9-cm petri dish).
3. Level dish and add 2 × 10^6 cells, 10^6/ml in 2 ml medium to top of filter, taking care not to perforate filter.
4. Place in humid CO_2 incubator in protective box. It is critical to avoid shaking the box, and the cultures should not be moved in the incubator to avoid spillage and resultant contamination.
5. Monolayer (or multilayer) should become established in about 5 d. It forms before then but complete integrity (formation of matrix, cell contacts, polarity, etc.) takes several days.
6. Culture may be maintained indefinitely, replacing medium every 3-5 d or transferring to fresh dish. A second cell type may be added on top if it is desired to study interactions, or to the other side of the filter by inverting the filter holder.

Analysis

1. *Penetration of cell through filter*. Trypsinize and count each side of filter in turn (Trypsinized cells not pass through even an 8-μm

filter as their spherical diameter in suspension exceeds this), or fix, embed, and section by electron microscope or conventional histology. Visualization is possible in whole mounts by mounting the fixed, stained (Giemsa) filter on a slide in DePeX under a coverslip under pressure to flatten it. Differential counting can then be performed by focusing on each plane alternately.

2. Detachment of cells from filter to bottom of dish. Count by trypsinization or scanning.
3. Partition above and below filter (chemotaxis or invasion) either by counting as in step 1 above or by prelabeling the cells with rhodamine or fluoresein isothiocyanate (5 μg/ml for 30 mins in trypsinized suspension) and measuring the fluorescence of solubilized cells (0.1% SDS in 0.3 N NaOH for 30 min) trypsinized from either side of the filter.
4. *Invasion.* Precoat filter with cell layer (normal fibroblasts, MDCK etc.), ensure confluence is achieved by microscopic examination, and then seed EDTA dissociated test cells on top of preformed layer (10^5–10^6 cells per filter). If test cells are RITC or FITC-labeled, fluorescent measurements will reveal appearance of cells below filter.

Variations

1. *Type of filter.* Millipore filters (preformed chambers) may be used. These are presterilized but opaque and of the mesh type filter matrix. Cells can be grown directly on the filter or after coating with gelatin or collagen.
2. *Filter porosity.* One-μm filters allow cell interaction and contact without transit of filter. Eight μm filters allow live cells to cross. Filters of 0.2 μm probably do not allow cell contact. Low-porosity filters may be used to study cell interaction without intermingling.
3. *Transfilter combination.* Invert filter and load underside first with 0.5 ml, 2 × 10^6/ml of cell suspension. After 2 d invert filter and load well as above.

19

CLONING OF CELLS

The ability of plant cells to de-differentiate into callus tissue in response to wounding has already been investigated. Callus cells placed in an artificial culture medium can be maintained in a persistent undifferentiated state. In this module, you will explore the ability of cultured plant cells to redifferentiate into whole plants. In cell culture, differentiation is effected by plant growth substances (plant hormones) in the culture medium. Cell culture allows us to control the concentration of plant growth regulators and thus affords some control over the differentiation of the plant cells. By changing the hormonal composition of the culture medium we can encourage the growth of roots, shoots, or callus. Two classes of plant growth regulators are used for this purpose, auxins and cytokinins. The specific plant growth regulators used in this experiment are indoleacetic acid (IAA, an *auxin*) and kinetin (a *cytokinin*). The media formulations to be tested contain different combinations of these components. As you perform the experiment, be mindful of the correlation between cytokinin/auxin and organ development.

Safety Guidelines

Follow standard laboratory safety practices.

The autoclave or pressure cooker required for media preparation is a burn hazard.

Alcohol or gas flames used to sterilize instruments during the lab class are extremely hazardous. Exercise due care to avoid igniting hair or clothing.

Experimental Outline

1. Prepare plant tissue

2. Make explants
3. Inoculate medium with explants
4. Observe cultures over a period of several weeks

Materials

Plant material

African violet (*Saintpaulia ionantha*) leaves. Choose leaf material that is not damaged, wilted, or bruised.

Culture media

MS (*Murashige* and *Skoog*) minimal organics medium to which four different combinations of plant growth regulators have been added:

		IAA (mg/l)	*Kinetin (mg/l)*
Medium	1	0	0
	2	0.1	10.0
	3	1.0	0
	4	10.0	5.0

Pre-lab Preparation

Timetable of events

Prepare explants (30 min.)

Inoculate media with tissue (30 min.)

Observe cultures over a period of several weeks

This module is an extension of the techniques and principles presented in "Establishing a Plant Cell Culture". In this experiment, cell cultures will be established on four different media, each containing a different combination of the plant growth regulators indoleacetic acid (IAA, an auxin) and kinetin (a cytokinin). In general, high cytokinin/auxin favors the formation of shoots, while low cytokinin/auxin favors root growth. While endogenous hormone levels in tissue explants can introduce some variability into the experiment, the protocol is reliable and produces dramatic results for the students. As an extension of Establishing a Plant Cell Culture activities, callus produced in that laboratory exercise can be used as an explant in this exercise.

Plant material

African violet (*Saintpaulia ionantha*). The plant is easy to obtain and familiar to the students. Choose a healthy plant with firm, unblemished leaves. If plants with variegated leaves are used, regenerated shoots will be of two types, green and albino. This oddity emphasizes that regenerants come from single cells and that the green and white sectors of the leaf have different plastid genotypes; the plant is a genetic chimaera. Of course albinos can only be maintained in culture, but sometimes they can be grown to quite a large size. Other plant material can be used for this experiment. The best of alternatives are probably tobacco (use leaf as explant) and cauliflower (use small florets as explant).

Media

Prepare MS minimal organics medium (with added sucrose and agar). Divide the medium into four aliquots. Add growth regulators from stock solutions according to the following schedule:

		IAA (mg/l)	*Kinetin (mg/l)*
Aliquot	1	0	0
	2	0.1	10.0
	3	1.0	0
	4	10.0	5.0

After adding the plant growth regulators, dispense the media into flasks or tubes for culture and autoclave.

Method

1. Prepare leaf tissue from African violet for cell culture as practiced in module 1. Briefly, disinfect leaf tissue by sequential washing, soaking in 70% ethanol, soaking in 10% household bleach, and rinsing in three changes of sterile distilled water.
2. Maintaining the sterility of the leaves, cut explants (about 1 cm^2) from the leaf blade or about 0.5 cm long from the petiole.
3. Distribute the explants among the four different media to be tested. Label the cultures with your name and date (include type of medium if this information is not already on the culture). Incubate your cultures under fluorescent lights (16 hr day, 26°C or room temperature).
4. Check the cultures at least once per week, observing the development of the cultures and looking for contamination by bacteria or fungi. Grossly contaminated cultures should be autoclaved before disposal. A small amount of contamination might be tolerated if it doesn't take over the culture.
5. The cultures should be allowed to develop for 4-6 weeks.

Results

Which of the cultures produced callus, roots, shoots? What cytokinin/auxin (kinetin/IAA) ratios favor shoot and root formation?

To continue the experiment, shoots can be excised and transferred to a medium with no hormones. This should encourage root growth. Rooted plants can eventually be transferred into soil if high humidity is maintained around the plant while it is becoming established. Cover the plant with a plastic bag or inverted jar to accomplish this. Explants that do not form shoots can be transferred to a medium that encourages shoot formation.

INDEX